Communications
in Computer and Information Science   844

Commenced Publication in 2007
Founding and Former Series Editors:
Phoebe Chen, Alfredo Cuzzocrea, Xiaoyong Du, Orhun Kara, Ting Liu,
Dominik Ślęzak, and Xiaokang Yang

More information about this series at http://www.springer.com/series/7899

Geetha Ganapathi · Arumugam Subramaniam
Manuel Graña · Suresh Balusamy
Rajamanickam Natarajan
Periakaruppan Ramanathan (Eds.)

Computational Intelligence, Cyber Security and Computational Models

Models and Techniques for Intelligent Systems and Automation

Third International Conference, ICC3 2017
Coimbatore, India, December 14–16, 2017
Proceedings

 Springer

Editors
Geetha Ganapathi
PSG College of Technology
Coimbatore
India

Arumugam Subramaniam
BITS Pilani, KK Birla
Goa
India

Manuel Graña
University of the Basque Country
San Sebastian
Spain

Suresh Balusamy
PSG College of Technology
Coimbatore
India

Rajamanickam Natarajan
PSG College of Technology
Coimbatore
India

Periakaruppan Ramanathan
PSG College of Technology
Coimbatore
India

ISSN 1865-0929   ISSN 1865-0937 (electronic)
Communications in Computer and Information Science
ISBN 978-981-13-0715-7   ISBN 978-981-13-0716-4 (eBook)
https://doi.org/10.1007/978-981-13-0716-4

Library of Congress Control Number: 2018944393

This Springer imprint is published by the registered company Springer Nature Singapore Pte Ltd.
The registered company address is: 152 Beach Road, #21-01/04 Gateway East, Singapore 189721, Singapore

Preface

Advances in computer hardware, software, and networking areas have provided ample scope for research and development in computer science theory and applications. In this connection, the Department of Applied Mathematics and Computational Sciences of PSG College of Technology organized the Third International Conference on Computational Intelligence, Cyber Security, and Computational Models (ICC3 2017) during December 14–16, 2017.

The theme "Models and Techniques for Intelligent Systems and Automation" was selected to promote high-quality research in the areas of computational models, computational intelligence, and cyber security and to emphasize the role of technology in real-time applications. These emerging interdisciplinary research areas have helped to solve multifaceted problems and gained prominent attention in recent years with the coming of the digital age.

Computational intelligence encompasses a broad set of techniques such as data mining, machine learning, deep learning, big data, the Internet of Things, and artificial intelligence. This track aims to bring out novel techniques for computation and visualization, find solutions for computationally expensive problems, and explore data within them be it classification, clustering, or feature engineering.

Cyber security has gained importance with the number of cyber threats escalating worldwide. There is a need for comprehensive security analysis, assessment, and actions to protect critical infrastructure and sensitive information. This track in the conference aims to bring together researchers, practitioners, developers, and users to explore cutting-edge ideas and end results in applied cryptography, intrusion detection and prevention, malware and botnets, security protocols, and system security.

Computational models have become extremely important for meeting the demands of the computer systems and for exploring new potential in the areas of topology, theory of computing, optimization techniques, database models, graph models, and stochastic models. Rigorous mathematical algorithms are of high demand for computational insight and necessitate vigorous tools for efficient implementation.

We received 64 papers in total and accepted 15 (24%). Each paper went through a stringent review process and where issues remained, additional reviews were commissioned. There were also an additional six papers from the plenary speakers related to their talk.

The organizers of ICC3 2017 wholeheartedly appreciate the efforts of the peer reviewers for their valuable comments in ensuring the quality of the proceedings. We extend our deepest gratitude to Springer for their continued support and for bringing out the proceedings on time with excellent production quality. We would like to thank all the keynote speakers, advisory committee members, and the session chairs for their timely contribution. We believe that the participants had fruitful research deliberations and benefitted by the excellent interaction.

ICC3 has been a biennial conference since 2013 and is successful in hosting a good dialogue between industry and academia. We hope that this tradition of academic interaction continues and fosters a greater binding in the years to come.

April 2018

Geetha Ganapathi
S. Arumugam
Manuel Graña
Suresh Balusamy
N. Rajamanickam
RM. Periakaruppan

Organization

Chief Patron

Shri. L. Gopalakrishnan   PSG & Sons Charities Trust, Coimbatore, India

Patron

R. Rudramoorthy   PSG College of Technology, Coimbatore, India

Organizing Chair

R. Nadarajan   PSG College of Technology, Coimbatore, India

Program Chairs

Geetha Ganapathi   PSG College of Technology, Coimbatore, India
S. Arumugam   BITS Pilani, KK Birla, Goa, India

Honorary Chairs

Andrew H. Sung   University of Southern Mississipi, USA
Dipankar Das Gupta   University of Memphis, USA

Computational Intelligence Track Chairs

Manuel Graña   University of the Basque Country, Spain
Suresh Balusamy   PSG College of Technology, Coimbatore, India

Cyber Security Track Chair

N. Rajamanickam   PSG College of Technology, Coimbatore, India

Computational Models Track Chair

RM. Periakaruppan   PSG College of Technology, Coimbatore, India

Advisory Committee

Rein Nobel   Vrije University, Amsterdam, The Netherlands
Kiseon Kim   Gwangju Institute of Science and Technology, South Korea
Phalguni Gupta   Indian Institute of Technology, Kanpur, India

R. Anitha	PSG College of Technology, India
G. Sai Sundara Krishnan	PSG College of Technology, India
R. S. Lekshmi	PSG College of Technology, India
M. Senthilkumar	PSG College of Technology, India
R. Vijayalakshmi	Miami University, USA

Technical Program Committee

Manual Graña	University of the Basque Country, Spain
Atilla Elci	Aksaray University, Turkey
Rein Nobel	VU University, Amsterdam, The Netherlands
Kiseon Kim	Gwangju Institute of Science and Technology, South Korea
Marten Wan Dijk	University of Connecticut Storrs, USA
Assaf Schuster	Israel Institute of Technology, Israel
Andrew H. Sung	University of Southern Mississipi, USA
Dipankar Das Gupta	University of Memphis, USA
Kurunathan Ratnavelu	University of Malaya, Malaysia
Bing Wu	Fayetteville University, USA
Nandagopal D.	University of South Australia, Australia
Michal Wozniak	Wroclaw University of Technology, Poland
S. Arumugam	BITS Pilani, KK Birla, Goa, India
R. Vijayalakshmi	Miami University, Ohio, USA
Nabaraj Dahal	Research Engineer elmTEK Pty. Ltd., Australia
Naga Dasari	Federation University Australia, Australia
Sridhar V.	Vencore Labs, USA
Aresh Dadlani	Nazarabayev University, Kazakhstan
N. P. Mahalik	Jordan College of Agricultural Sciences and Technology, Jordan
Srinivas R. Chakravarthy	Kettering University, USA
Hironori Washizaki	Waseda University, Japan
Giovanni Semeraro	University of Bari Aldo Moro, Italy
A. Somasundaram	BITS Pilani, Dubai Campus, Dubai
S. Arockiasamy	University of Nizwa, Sultanate of Oman
Natarajan Meghanathan	Jackson State University, USA
Eduardo B. Fernandez	Florida Atlantic University, USA
Carlos R. Carpintero F.	Universidad de Oriente, Venezuela
Hironori Washizaki	Waseda University, Japan
Igor Kotenko	St. Petersburg Institute for Informatics and Automation, Russia
Douglas S. Reeves	NC State University, USA
Janet Light	University of New Brunswick, Canada
Subir Kumar Ghosh	Ramakrishna Mission Vivekananda University, India
Phalguni Gupta	IIT Kanpur, India
C. Pandurangan	IIT Madras, India
B. Raveendran	IIT Madras, India

Vineeth N. Balasubramanian	IIT Hyderabad, India
Sudip Misra	IIT Kharagpur, India
M. Ashok Kumar	IIT Indore, India
K. Viswanathan Iyer	NIT Trichy, India
U. Dinesh Kumar	IIM Bangalore, India
C. Mala	NIT Trichy, India
Murugesan	NIT Trichy, India
N. P. Gopalan	NIT Trichy, India
S. K. Hafizul Islam	IIIT Kalyani, India
Rishi Ranjan Singh	IIIT Allahabad, India
Chandrasekaran	NIT Surathkal, India
Vijay Bhaskar Semwal	IIIT Dharwad, India
R. Nadarajan	PSG College of Technology, Coimbatore, India
A. Kandasamy	NIT Surathkal, India
R. B. V. Subramanyam	NIT Warangal, India
Manoj Changat	University of Kerala, India
Sethumadhavan M.	Amrita University, Coimbatore, India
Somasundaram K.	Amrita University, Coimbatore, India
R. Anitha	PSG College of Technology, Coimbatore, India
Latha Parameswaran	Amrita University, Coimbatore, India
G. Sethuraman	Anna University, Chennai, India
J. Baskar Babujee	Anna University, Chennai, India
G. Sai Sundara Krishnan	PSG College of Technology, Coimbatore, India
Ruhul Amin	Thapar University, Patiala, India
K. S. Sridharan	Sri Sathya Sai Institute of Higher Learning, (Deemed University), India
Geetha Ganapathi	PSG College of Technology, Coimbatore, India
Suresh C. Satapathy	ANITS, Vishakapatnam, India
R. S. Lekshmi	PSG College of Technology, Coimbatore, India
S. Indumathi	Ambedkar Institute of Technology, Bengaluru, India
RM. Periakaruppan	PSG College of Technology, Coimbatore, India
C. R. Subramanian	Institute of Mathematical Sciences, Chennai, India
M. Senthilkumar	PSG College of Technology, Coimbatore, India
R. Ramanujam	Institute of Mathematical Sciences, Chennai, India
Suresh Balusamy	PSG College of Technology, Coimbatore, India
Shina Sheen	PSG College of Technology, Coimbatore, India
K. Thangavel	Periyar University, Salem, India
R. Venkatesan	PSG College of Technology, Coimbatore, India
T. Purushothaman	Government College of Technology, Coimbatore, India
G. Sudha Sadasivam	PSG College of Technology, Coimbatore, India
Elizabeth Sherly	IIITM – Trivandrum, India
A. Chitra	PSG College of Technology, Coimbatore, India
G. Padmavathi	Avinashilingam Deemed University, Coimbatore, India
P. Navaneethan	PSG College of Technology, Coimbatore, India
H. S. Ramane	Karnatak University, Dharwad, India

Tarkeshwar Singh   BITS, Pilani, Goa, India
B. Sivakumar   Madurai Kamaraj University, India
P. Balasubramaniam   Gandhigram Rural University, India

Sponsors

Cognizant Technology Solutions, Chennai, India

Science and Engineering Research Board, Department of Science and Technology, Government of India

Contents

Computational Models

Computational Intelligence

A Short Review of Recent ELM Applications

Manuel Graña[✉]

Computational Intelligence Group, University of the Basque Country,
Leioa, Spain
manuel.grana@ehu.es

Abstract. Extreme Learning Machine has enjoyed an explosive growth in the last years. Starting from a "simple" idea of replacing the slow backpropagation training of the input to hidden units weights by a random sampling, they have been enriched and hybridized in many ways, becoming an accepted tool in the machine learning engineer toolbox. Hence their multiple applications, which are growing in the last few years. The aim of this short review is to gather some glimpses on the state of development and application of ELM. The point of view of the review is factual: ELM are happening, in the sense that many researchers are using them and finding them useful in a wide diversity of applications.

1 Introduction

Randomization in learning systems is not new to Extreme Learning Machines (ELM) [31]. The Random Forest proposed originally by Breiman [13] is probably the foremost example. The idea is to build a collection of decision trees picking features and splitting points at random, and surprisingly it works!. Recent reviews [28,29,56] cover fundamental aspects such as the extension of ELM from classification to clustering, representation learning and others, as well as the spectrum of applications. This review gathers more recent contributions in the literature. The main emphasis is in applications, which show the real usefulness of the approach.

2 Architectures and Hybrids

There are a continuous effort in the community to propose enhancements of ELM, such as innovative weight computation techniques allowing two hidden layers [57], hierarchical multilayer structures [83], local connectivity including subgraphs [30], dynamic adjustment of hidden node parameters [22], or the optimal pruning improved by proposed acceleration of the singular value decomposition by [26], improvements in the application of the kernel trick, such as the multikernel algorithms [23], the application of matrix decomposition techniques to speed up output weights estimation in large networks [42] and inverse-free training techniques [43], the proposal of new regularization techniques [48] revisiting

© Springer Nature Singapore Pte Ltd. 2018
G. Ganapathi et al. (Eds.): ICC3 2017, CCIS 844, pp. 3–12, 2018.
https://doi.org/10.1007/978-981-13-0716-4_1

the classical L1 and L2 norm regularization procedures, or the reduced kernel approach [21] which has also been extended to online learning [20]. Also, the online sequential ELM continues to be updated and improved, either by parallel implementations [32], use of regularization [60], adding adaptive kernel training [59], or by new features such as semisupervised learning [38]. Evolutionary algorithms are successfully applied to select optimal ELM model for the task at hand [40]. Also, nanotechnological implementations are possible based on the randomness of deposition in nanoscale filamentary-resistive memory devices implementing the hidden layer weights [69] and nonvolatile domain-wall nanowires [80] achieving high energy efficiencies.

The use of ensemble approaches is a very positive move to increase the stability of the ELM results [14]. By naive application of the Law of Large Numbers, the ensemble converges to zero error as the number of individual classifiers/regressors increases. For instance, ELM can be very effectively embedded in heterogeneous ensembles where they play a specific role [3,4]. On other approaches, ensembles of ELM serve to build probabilistic threshold querying [41] to deal with situations where the data uncertainty is high and we prefer to have a multiple object response. Ordered aggregation ensembles are able to adapt quickly to non-stationary environments [63]. It is also apparent that innovative propositions of activation functions, such as the q-Gaussian [67], or the generalized polynomial [76] lead to improved ensemble performance. Data partition allows multiELM to achieve improved performance [84].

Most interestingly, recent trends involve ELMs in some kind of hybridization scheme, either enhancing them or profitting from its quick learning. Some authors have used a manifold learning approach [77] for the regularization of the ELM ensuring that it is coherent with the local tangent space of the data, or for the clustering of data prior to the ELM classification [55], or using a supervised discriminant manifold approach [53]. Others have enforced a graph regularization based on the hidden node outputs in order to enforce their similarity for data from the same class [54], or simply using conventional graph embedding techniques for feature extraction [34]. Another hybridization examples come from the use of statistical techniques, such as the use of robust M-estimation in ELM training to avoid outlier effects which are caused by mislabeled samples [8,24], the use of inductive bias (supervised learning conditioned to unsupervised learning results) to achieve semi-supervised learning in ELM [12], the use of sparse bayesian estimation to improve regression in the case of highly heterogeneous clustered data [39], the application of Gaussian Mixture Models for missing data distribution estimation prior to ELM multiple imputation [66], or the statistical characterization of uncertain data leading to new ELM approaches to deal with them [15]. SemiRandom Projections lead to the formulation of partially connected ELM (PC-ELM) which achieves remarkable results in document processing [86]. The hybridization with wavelet signal analysis has lead to the formulation of specific dual activation functions in the summation wavelet ELM [37] which improves processing of dynamic time series data.

3 Applications

3.1 Control

Control is a very fruitful field of application of the ELMs because of their quick learning, that makes them specially suitable to model plant dynamics for the control loop. They have been applied for instance to: model the dynamics of homogeneous charge compression ignition (HCCI) engines [35] for predictive control loops. Similarly, a wavelet transform based ELM-WT serves to model the exergetic profile of diesel engines using blended combustible [1]. In the case of electrical power operation systems, ELM have been used to model the Critical Clearing Time (CCT) [68] which is the time allowed to remove a perturbation in the system with affecting operation. But they can also be used to build the actual control system, such as the bi-level kernel ELM used to control the cold-start emissions of engines [7]. A chip curing process is a heating process that is time varying and highly non-linear, which has been modeled by ELM endowed with adaptive online regularization factor [47]. In other works [70,81], NOX emisions of a coal boiling plant are modeled by ELM allowing the optimal control in real time by harmony search methods. The approach is also followed for the development of a model of the dynamic operation of homogeneous charge compression ignition engines [36], a technology that promises great impact in the automotive industry but that suffers from extreme non-linearities.

The ELMs are also used as surrogate for an unknown dynamical model of a complicated multi-robot system carrying a linear flexible object, i.e. a hose, to some destination. The surrogate ELM is used by a reinforcement learning approach to learn the control of the robots [45]. Another surrogate model of the environment has been proposed for unmanned vehicle control [75], based on an error surface built on the fly. Yet another of this kind models the nonlinearities of rotor position tracking for active magnetic bearing [85] in order to apply an adaptive backstepping neural controller.

Soft sensors in chemical processes are computational predictions of key variables that are unobservable, and whose values can not be derived analytically. A variation of ELM incorporating partial least squares and robust bagging are able soft sensors in complex chemical processes [27]. Sensing the mixture of pulverized biomass and coal in cofired energy power plants is also achieved by adaptive wavelet ELM [78]. Quality control in batchs forging processes is achieved by online probabilistic ELM distribution modeling [46]. Control of water level in U-Tube steam generation systems is also achieved by ELM after autoregressive (NARX) modeling [11].

3.2 Remote Sensing

ELMs have been applied to classification of hyperspectral images in order to obtain thematic maps of cultivated lands [51] for agricultural and land use control, in some works evolutionary algorithms such as differential evolution are used for ELM model selection [9]. Ensemble models have been proposed to tacke with hyperspectral data [58]. They have bee also combined with spatial regularization to achieve correction of false class assignment [6]. Other works combine random subspace selection with ELM [82] in an approach that models spatial information by extended multiattribute profiles. Texture features in the form of local binary patterns have been used for such spatial regularization [44].

Besides classification, ELM regression can be used for the computation of endmember abundances in spectral unmixing processes [5]. Dynamic ensemble selection achieves optimal configuration by selecting the best classifiers local to each pixel [19]. Smart image reconstruction for cloud removal is highly desirable to avoid discarding large chunks of optical remote sensing data [16], ELM improve over other architectures. Working on the JPEG2000 coefficients of pancromatic optical satellite images, a combination of ELM and deep learning has been able to perform ship detection in ocean scenes [71]. Classification of general satellite images has been achieved by fusion of ELM and graph optimization for the regularization of classification results [10]. Domain adaptation is of increasing importance in remote sensing, because it allows efficient processing of time series of images, by small effort adaptation of classifiers trained in past images. Three layer ELM has been successfully developed for this in VHR image [52].

3.3   Neuroimage and Neuroscience

Most efforts are directed to neuroimage processing to obtain image biomarkers with a handful of signal capture methods, where the star are the diverse modalities of magnetic resonance imaging (MRI). Detection of schizophrenia from resting state functional magnetic resonance imaging (fMRI) has been achieved by ensembles of ELM [18] over local activity measures. Alzheimer's disease morphological differences relative to healthy subjects have been detected using ELM as the wrapper classifier to evaluate the discriminant power of brain regions [72]. Besides, ELM has been applied to electroencephalography data (EEG) for several tasks, such as emotion recognition [53,55], and to epileptic seizure detection based on a innovative feature fusion approach [64], and entropy-based features in an intracraneal EEG [65]. A chipset for EEG recognition implements ELM as the classification after feature extraction [17] for Brain Machine Interfaces.

3.4   Image Processing and Face Recognition

Approaches to image classification follow classical paths, like the feature extraction by some transformation, for instance obtained from manifold learning [55]. Recently, the deep learning architecture built by ELM autoencoders has been applied to benchmarking image recogntion tasks [73], and to the extraction of 3D features [79]. High resolution imaging in lab-on-chip microfluid inspection for the detection of specific cells in cultives requires super-resolution approaches that have been achieved by ELM [33] in a single frame. Image content can be used for website categorization by classification with ELM [2].

Face recognition and related topics, such as emotion recognition, has been a key benchmarking problem in artificial intelligence for many years. It has been used also to landmark the progress of ELM research. Face recognition by ELM approaches has been demonstrated with standard benchmark databases [49,50] such as the ORL, Yale and CMU databases. A common pipeline is the realization of some dimensionality reduction (PCA, LICA, B2DPCA) prior to enter the image to the classification. They have been accepted to the point of being proposed for cloud based image labeling [74]. A special kind of face information that is being looked upon are the face expressions, aiming to detect its the emotional content. Adaboost with ELM as the base classifier [25] use specific features based on self-similar gradients of the image for the detection of smiles in face images. Sparse representation is used for face feature extraction in [61] aiming to the recognition of emotion from face expressions. Watermarking images by embedding data in image transformation coefficients can be very fragile to image transformations and noise. ELM have been succesfully applied to recover altered watermarks in the wavelet transform domain [62].

4   Conclusions

ELMs have a very dynamic ecosystem for the development of new architectures, which extremely fruitful in applications. Seems that they have been accepted in the toolbox of the data analitics engineer and has been combined and hybridized in many ways, which attests to the approach flexibility. Looks like they have come to stay.

References

1. Aghbashlo, M., Shamshirband, S., Tabatabaei, M., Yee, P.L., Larimi, Y.N.: The use of ELM-WT (extreme learning machine with wavelet transform algorithm) to predict exergetic performance of a DI diesel engine running on diesel/biodiesel blends containing polymer waste. Energy **94**, 443–456 (2016)
2. Akusok, A., Miche, Y., Karhunen, J., Bjork, K.-M., Nian, R., Lendasse, A.: Arbitrary category classification of websites based on image content. IEEE Comput. Intell. Mag. **10**(2), 30–41 (2015)
3. Ayerdi, B., Romay, M.G.: Hyperspectral image analysis by spectral-spatial processing and anticipative hybrid extreme rotation forest classification. IEEE Trans. Geosci. Remote Sens. **PP**(99), 1–13 (2015)
4. Ayerdi, B., Graña, M.: Hybrid extreme rotation forest. Neural Netw. **52**, 33–42 (2014)
5. Ayerdi, B., Graña, M.: Hyperspectral image nonlinear unmixing and reconstruction by ELM regression ensemble. Neurocomputing **174**(Part A), 299–309 (2016)
6. Ayerdi, B., Marqués, I., Graña, M.: Spatially regularized semisupervised ensembles of extreme learning machines for hyperspectral image segmentation. Neurocomputing **149**(Part A), 373–386 (2015). Advances in Extreme Learning Machines
7. Azad, N.L., Mozaffari, A., Hedrick, J.K.: A hybrid switching predictive controller based on bi-level kernel-based ELM and online trajectory builder for automotive coldstart emissions reduction. Neurocomputing **173**(Part 3), 1124–1141 (2016)

8. Barreto, G.A., Barros, A.L.B.P.: A robust extreme learning machine for pattern classification with outliers. Neurocomputing **176**, 3–13 (2016). Recent Advancements in Hybrid Artificial Intelligence Systems and Its Application to Real-World Problems Selected Papers from the HAIS 2013 Conference

9. Bazi, Y., Alajlan, N., Melgani, F., AlHichri, H., Malek, S., Yager, R.R.: Differential evolution extreme learning machine for the classification of hyperspectral images. Geosci. Remote Sens. Lett. IEEE **11**(6), 1066–1070 (2014)

10. Bencherif, M.A., Bazi, Y., Guessoum, A., Alajlan, N., Melgani, F., AlHichri, H.: Fusion of extreme learning machine and graph-based optimization methods for active classification of remote sensing images. Geosci. Remote Sens. Lett. IEEE **12**(3), 527–531 (2015)

11. Beyhan, S., Kavaklioglu, K.: Comprehensive modeling of U-tube steam generators using extreme learning machines. IEEE Trans. Nucl. Sci. **62**(5), 2245–2254 (2015)

12. Bisio, F., Decherchi, S., Gastaldo, P., Zunino, R.: Inductive bias for semi-supervised extreme learning machine. Neurocomputing **174**(Part A), 154–167 (2016)

13. Breiman, L.: Random forests. Mach. Learn. **45**(1), 5–32 (2001)

14. Cao, J., Lin, Z., Huang, G.-B., Liu, N.: Voting based extreme learning machine. Inf. Sci. **185**(1), 66–77 (2012)

15. Cao, K., Wang, G., Han, D., Bai, M., Li, S.: An algorithm for classification over uncertain data based on extreme learning machine. Neurocomputing **174**(Part A), 194–202 (2016)

16. Chang, N.-B., Bai, K., Chen, C.-F.: Smart information reconstruction via time-space-spectrum continuum for cloud removal in satellite images. IEEE J. Sel. Top. Appl. Earth Obs. Remote Sens. **8**(5), 1898–1912 (2015)

17. Chen, Y., Yao, E., Basu, A.: A 128-channel extreme learning machine-based neural decoder for brain machine interfaces. IEEE Trans. Biomed. Circ. Syst. **PP**(99), 1 (2015)

18. Chyzhyk, D., Savio, A., Graña, M.: Computer aided diagnosis of schizophrenia on resting state fMRI data by ensembles of ELM. Neural Netw. **68**, 23–33 (2015)

19. Damodaran, B.B., Nidamanuri, R.R., Tarabalka, Y.: Dynamic ensemble selection approach for hyperspectral image classification with joint spectral and spatial information. IEEE J. Sel. Top. Appl. Earth Obs. Remote Sens. **8**(6), 2405–2417 (2015)

20. Deng, W.-Y., Ong, Y.-S., Tan, P.S., Zheng, Q.-H.: Online sequential reduced kernel extreme learning machine. Neurocomputing **174**(Part A), 72–84 (2016)

21. Deng, W.-Y., Ong, Y.-S., Zheng, Q.-H.: A fast reduced kernel extreme learning machine. Neural Netw. **76**, 29–38 (2016)

22. Feng, G., Lan, Y., Zhang, X., Qian, Z.: Dynamic adjustment of hidden node parameters for extreme learning machine. IEEE Trans. Cybern. **45**(2), 279–288 (2015)

23. Fossaceca, J.M., Mazzuchi, T.A., Sarkani, S.: MARK-ELM: application of a novel multiple kernel learning framework for improving the robustness of network intrusion detection. Expert Syst. Appl. **42**(8), 4062–4080 (2015)

24. Frenay, B., Verleysen, M.: Reinforced extreme learning machines for fast robust regression in the presence of outliers. IEEE Trans. Cybern. **PP**(99), 1–13 (2015)

25. Gao, X., Chen, Z., Tang, S., Zhang, Y., Li, J.: Adaptive weighted imbalance learning with application to abnormal activity recognition. Neurocomputing **173**(Part 3), 1927–1935 (2016)

26. Grigorievskiy, A., Miche, Y., Käpylä, M., Lendasse, A.: Singular value decomposition update and its application to (inc)-OP-ELM. Neurocomputing **174**(Part A), 99–108 (2016)

27. He, Y.L., Geng, Z.Q., Zhu, Q.X.: Soft sensor development for the key variables of complex chemical processes using a novel robust bagging nonlinear model integrating improved extreme learning machine with partial least square. Chemom. Intell. Lab. Syst. **151**, 78–88 (2016)
28. Huang, G., Cambria, E., Toh, K., Widrow, B., Xu, Z.: New trends of learning in computational intelligence [guest editorial]. IEEE Comput. Intell. Mag. **10**(2), 16–17 (2015)
29. Huang, G., Huang, G.-B., Song, S., You, K.: Trends in extreme learning machines: a review. Neural Netw. **61**, 32–48 (2015)
30. Huang, G.-B., Bai, Z., Kasun, L.L.C., Vong, C.M.: Local receptive fields based extreme learning machine. IEEE Comput. Intell. Mag. **10**(2), 18–29 (2015)
31. Huang, G.-B., Zhu, Q.-Y., Siew, C.-K.: Extreme learning machine: theory and applications. Neurocomputing **70**(1–3), 489–501 (2006). Neural Networks Selected Papers from the 7th Brazilian Symposium on Neural Networks (SBRN 2004) 7th Brazilian Symposium on Neural Networks
32. Huang, S., Wang, B., Qiu, J., Yao, J., Wang, G., Yu, G.: Parallel ensemble of online sequential extreme learning machine based on mapreduce. Neurocomputing **174**(Part A), 352–367 (2016)
33. Huang, X., Yu, H., Liu, X., Jiang, Y., Yan, M.: A single-frame superresolution algorithm for lab-on-a-chip lensless microfluidic imaging. IEEE Des. Test **32**(6), 32–40 (2015)
34. Iosifidis, A., Tefas, A., Pitas, I.: Graph embedded extreme learning machine. IEEE Trans. Cybern. **46**(1), 311–324 (2016)
35. Janakiraman, V.M., Nguyen, X., Assanis, D.: An ELM based predictive control method for HCCI engines. Eng. Appl. Artif. Intell. **48**, 106–118 (2016)
36. Janakiraman, V.M., Nguyen, X., Sterniak, J., Assanis, D.: Identification of the dynamic operating envelope of HCCI engines using class imbalance learning. IEEE Trans. Neural Netw. Learn. Syst. **26**(1), 98–112 (2015)
37. Javed, K., Gouriveau, R., Zerhouni, N.: SW-ELM: a summation wavelet extreme learning machine algorithm with a priori parameter initialization. Neurocomputing **123**, 299–307 (2014). Contains Special Issue Articles: Advances in Pattern Recognition Applications and Methods
38. Jia, X., Wang, R., Liu, J., Powers, D.M.W.: A semi-supervised online sequential extreme learning machine method. Neurocomputing **174**(Part A), 168–178 (2016)
39. Kiaee, E., Sheikhzadeh, H., Mahabadi, S.E.: Sparse Bayesian mixed-effects extreme learning machine, an approach for unobserved clustered heterogeneity. Neurocomputing **175**(Part A), 411–420 (2016)
40. Lacruz, B., Lahoz, D., Mateo, P.M.: μG2-ELM: an upgraded implementation of μG-ELM. Neurocomputing **171**, 1302–1312 (2016)
41. Li, J., Wang, B., Wang, G., Zhang, Y.: Probabilistic threshold query optimization based on threshold classification using ELM for uncertain data. Neurocomputing **174**(Part A), 211–219 (2016)
42. Li, J., Hua, C., Tang, Y., Guan, X.: A fast training algorithm for extreme learning machine based on matrix decomposition. Neurocomputing **173**(Part 3), 1951–1958 (2016)
43. Li, S., You, Z.-H., Guo, H., Luo, X., Zhao, Z.-Q.: Inverse-free extreme learning machine with optimal information updating. IEEE Trans. Cybern. **PP**(99), 1 (2015)
44. Li, W., Chen, C., Hongjun, S., Qian, D.: Local binary patterns and extreme learning machine for hyperspectral imagery classification. IEEE Trans. Geosci. Remote Sens. **53**(7), 3681–3693 (2015)

45. Lopez-Guede, J.M., Fernandez-Gauna, B., Ramos-Hernanz, J.A.: A L-MCRS dynamics approximation by ELM for reinforcement learning. Neurocomputing **150**(Part A), 116–123 (2015)
46. Xinjiang, L., Liu, C., Huang, M.: Online probabilistic extreme learning machine for distribution modeling of complex batch forging processes. IEEE Trans. Ind. Inf. **11**(6), 1277–1286 (2015)
47. Lu, X., Zhou, C., Huang, M., Lv, W.: Regularized online sequential extreme learning machine with adaptive regulation factor for time-varying nonlinear system. Neurocomputing **174**(Part B), 617–626 (2016)
48. Luo, X., Chang, X., Ban, X.: Regression and classification using extreme learning machine based on L1-norm and L2-norm. Neurocomputing **174**(Part A), 179–186 (2016)
49. Marques, I., Graña, M.: Face recognition with lattice independent component analysis and extreme learning machines. Soft. Comput. **16**(9), 1525–1537 (2012)
50. Mohammed, A.A., Minhas, R., Wu, Q.M.J., Sid-Ahmed, M.A.: Human face recognition based on multidimensional PCA and extreme learning machine. Pattern Recogn. **44**(10–11), 2588–2597 (2011). Semi-Supervised Learning for Visual Content Analysis and Understanding
51. Moreno, R., Corona, F., Lendasse, A., Graña, M., Galvão, L.S.: Extreme learning machines for soybean classification in remote sensing hyperspectral images. Neurocomputing **128**, 207–216 (2014)
52. Othman, E., Bazi, Y., Alajlan, N., AlHichri, H., Melgani, F.: Three-layer convex network for domain adaptation in multitemporal VHR images. IEEE Geosci. Remote Sens. Lett. **PP**(99), 1–5 (2016)
53. Peng, Y., Lu, B.-L.: Discriminative manifold extreme learning machine and applications to image and EEG signal classification. Neurocomputing **174**(Part A), 265–277 (2016)
54. Peng, Y., Wang, S., Long, X., Lu, B.-L.: Discriminative graph regularized extreme learning machine and its application to face recognition. Neurocomputing **149**(Part A), 340–353 (2015). Advances in Extreme Learning Machines
55. Peng, Y., Zheng, W.-L., Lu, B.-L.: An unsupervised discriminative extreme learning machine and its applications to data clustering. Neurocomputing **174**(Part A), 250–264 (2016)
56. Principe, J.C., Chen, B.: Universal approximation with convex optimization: gimmick or reality? [discussion forum]. IEEE Comput. Intell. Mag. **10**(2), 68–77 (2015)
57. Qu, B.Y., Lang, B.F., Liang, J.J., Qin, A.K., Crisalle, O.D.: Two-hidden-layer extreme learning machine for regression and classification. Neurocomputing **175**(Part A), 826–834 (2016)
58. Samat, A., Du, P., Liu, S., Li, J., Cheng, L.: E^2 LMs: ensemble extreme learning machines for hyperspectral image classification. IEEE J. Sel. Top. Appl. Earth Obs. Remote Sens. **7**(4), 1060–1069 (2014)
59. Scardapane, S., Comminiello, D., Scarpiniti, M., Uncini, A.: Online sequential extreme learning machine with kernels. IEEE Trans. Neural Netw. Learn. Syst. **26**(9), 2214–2220 (2015)
60. Shao, Z., Er, M.J.: An online sequential learning algorithm for regularized extreme learning machine. Neurocomputing **173**(Part 3), 778–788 (2016)
61. Shojaeilangari, S., Yau, W.-Y., Nandakumar, K., Li, J., Teoh, E.K.: Robust representation and recognition of facial emotions using extreme sparse learning. IEEE Trans. Image Process. **24**(7), 2140–2152 (2015)

62. Singh, R.P., Dabas, N., Chaudhary, V., Nagendra: Online sequential extreme learning machine for watermarking in DWT domain. Neurocomputing **174**(Part A), 238–249 (2016)
63. Soares, S.G., Araújo, R.: An adaptive ensemble of on-line extreme learning machines with variable forgetting factor for dynamic system prediction. Neurocomputing **171**, 693–707 (2016)
64. Song, J.-L., Hu, W., Zhang, R.: Automated detection of epileptic EEGs using a novel fusion feature and extreme learning machine. Neurocomputing **175**(Part A), 383–391 (2016)
65. Song, Y., Zhang, J.: Discriminating preictal and interictal brain states in intracranial EEG by sample entropy and extreme learning machine. J. Neurosci. Methods **257**, 45–54 (2016)
66. Sovilj, D., Eirola, E., Miche, Y., Björk, K.-M., Nian, R., Akusok, A., Lendasse, A.: Extreme learning machine for missing data using multiple imputations. Neurocomputing **174**(Part A), 220–231 (2016)
67. Stosic, D., Stosic, D., Ludermir, T.: Voting based q-generalized extreme learning machine. Neurocomputing **174**(Part B), 1021–1030 (2016)
68. Sulistiawati, I.B., Priyadi, A., Qudsi, O.A., Soeprijanto, A., Yorino, N.: Critical clearing time prediction within various loads for transient stability assessment by means of the extreme learning machine method. Int. J. Electr. Power Energ. Syst. **77**, 345–352 (2016)
69. Suri, M., Parmar, V.: Exploiting intrinsic variability of filamentary resistive memory for extreme learning machine architectures. IEEE Trans. Nanotechnol. **14**(6), 963–968 (2015)
70. Tan, P., Xia, J., Zhang, C., Fang, Q., Chen, G.: Modeling and reduction of NOX emissions for a 700 mw coal-fired boiler with the advanced machine learning method. Energy **94**, 672–679 (2016)
71. Tang, J., Deng, C., Huang, G.-B., Zhao, B.: Compressed-domain ship detection on spaceborne optical image using deep neural network and extreme learning machine. IEEE Trans. Geosci. Remote Sens. **53**(3), 1174–1185 (2015)
72. Termenon, M., Graña, M., Savio, A., Akusok, A., Miche, Y., Björk, K.-M., Lendasse, A.: Brain MRI morphological patterns extraction tool based on extreme learning machine and majority vote classification. Neurocomputing **174**(Part A), 344–351 (2016)
73. Tissera, M.D., McDonnell, M.D.: Deep extreme learning machines: supervised autoencoding architecture for classification. Neurocomputing **174**(Part A), 42–49 (2016)
74. Vinay, A., Shekhar, V.S., Rituparna, J., Aggrawal, T., Murthy, K.N.B., Natarajan, S.: Cloud based big data analytics framework for face recognition in social networks using machine learning. Procedia Comput. Sci. Big Data, Cloud Comput. Challenges **50**, 623–630 (2015)
75. Wang, N., Sun, J.-C., Er, M.J., Liu, Y.-C.: A novel extreme learning control framework of unmanned surface vehicles. IEEE Trans. Cybern. **PP**(99), 1 (2015)
76. Wang, N., Er, M.J., Han, M.: Generalized single-hidden layer feedforward networks for regression problems. IEEE Trans. Neural Netw. Learn. Syst. **26**(6), 1161–1176 (2015)
77. Wang, Q., Wang, W., Nian, R., He, B., Shen, Y., Björk, K.-M., Lendasse, A.: Manifold learning in local tangent space via extreme learning machine. Neurocomputing **174**(Part A), 18–30 (2016)

78. Wang, X., Hongli, H., Liu, X.: Multisensor data fusion techniques with ELM for pulverized-fuel flow concentration measurement in cofired power plant. IEEE Trans. Instrum. Measur. **64**(10), 2769–2780 (2015)

79. Wang, Y., Xie, Z., Xu, K., Dou, Y., Lei, Y.: An efficient and effective convolutional auto-encoder extreme learning machine network for 3d feature learning. Neurocomputing **174**(Part B), 988–998 (2016)

80. Wang, Y., Hao, Y., Ni, L., Huang, G.-B., Yan, M., Weng, C., Yang, W., Zhao, J.: An energy-efficient nonvolatile in-memory computing architecture for extreme learning machine by domain-wall nanowire devices. IEEE Trans. Nanotechnol. **14**(6), 998–1012 (2015)

81. Wong, S.Y., Yap, K.S., Yap, H.J.: A constrained optimization based extreme learning machine for noisy data regression. Neurocomputing **171**, 1431–1443 (2016)

82. Xia, J., Mura, M.D., Chanussot, J., Du, P., He, X.: Random subspace ensembles for hyperspectral image classification with extended morphological attribute profiles. IEEE Trans. Geosci. Remote Sens. **53**(9), 4768–4786 (2015)

83. Yang, Y., Wu, Q.M.J.: Multilayer extreme learning machine with subnetwork nodes for representation learning. IEEE Trans. Cybern. **PP**(99), 1–14 (2015)

84. Yang, Y., Wu, Q.M.J., Wang, Y., Zeeshan, K.M., Lin, X., Yuan, X.: Data partition learning with multiple extreme learning machines. IEEE Trans. Cybern. **45**(8), 1463–1475 (2015)

85. Yang, Z.-X., Zhao, G.-S., Rong, H.-J., Yang, J.: Adaptive backstepping control for magnetic bearing system via feedforward networks with random hidden nodes. Neurocomputing **174**(Part A), 109–120 (2016)

86. Zhao, R., Mao, K.: Semi-random projection for dimensionality reduction and extreme learning machine in high-dimensional space. IEEE Comput. Intell. Mag. **10**(3), 30–41 (2015)

Critical Feature Selection and Critical Sampling for Data Mining

Bernardete Ribeiro[1], José Silva[1], Andrew H. Sung[2(✉)],
and Divya Suryakumar[3]

[1] Department of Informatics Engineering, University of Coimbra,
3030-290 Coimbra, Portugal
bribeiro@dei.uc.pt, jmpsilva@student.dei.uc.pt
[2] School of Computing, The University of Southern Mississippi,
Hattiesburg, MS 39406, USA
andrew.sung@usm.edu
[3] ConstructConnect, Cincinnati, USA
dsuryakumar@gmail.com

Abstract. The rapidly growing big data generated by connected sensors, devices, the web and social network platforms, etc., have stimulated the advancement of data science, which holds tremendous potential for problem solving in various domains. How to properly utilize the data in model building to obtain accurate analytics and knowledge discovery is a topic of great importance in data mining, and wherefore two issues arise: how to select a critical subset of features and how to select a critical subset of data points for sampling. This paper presents ongoing research that suggests: 1. the critical feature dimension problem is theoretically intractable, but simple heuristic methods may well be sufficient for practical purposes; 2. there are big data analytic problems where evidence suggest that the success of data mining depends more on the critical feature dimension than the specific features selected, thus a random selection of the features based on the dataset's critical feature dimension will prove sufficient; and 3. The problem of critical sampling has the same intractable complexity as critical feature dimension, but again simple heuristic methods may well be practicable in most applications; experimental results with several versions of the heuristic method are presented and discussed. Finally, a set of metrics for data quality is proposed based on the concepts of critical features and critical sampling.

Keywords: Data mining · Critical feature selection
Critical Sampling · Data quality

1 Introduction

One of the many challenges in utilizing "big data" is how to reduce the size of datasets in tasks such as data mining for model building or knowledge discovery. In that regard, effective feature ranking and selection algorithms may guide

© Springer Nature Singapore Pte Ltd. 2018
G. Ganapathi et al. (Eds.): ICC3 2017, CCIS 844, pp. 13–24, 2018.
https://doi.org/10.1007/978-981-13-0716-4_2

us in data reduction by eliminating features that are insignificant, dependent, noisy, unreliable, or irrelevant. In some bio or medical informatics datasets, for example, the number of features can reach tens of thousands. This is partly because that many datasets being constructed today for future data mining purposes, without prior knowledge about the relationships among features or specific objective about what is to be explored or derived from the data, likely have included measurable attributes that are actually insignificant or irrelevant, which inevitably results in large numbers of useless features that can be deleted to reduce the size of datasets without negative consequences in data analytics or data mining [1,5].

We investigate in this paper the general question: Given a dataset with p features, is there a Critical Feature Dimension (CFD, or the smallest number of features that are necessary) that is required, say, for a particular data mining or machine learning process, to satisfy a minimal performance threshold? This is a useful question to consider since feature selection methods generally provide no guidance on the number of features to include for a particular task; moreover, for many poorly-understood and complex problems to which big data brings some hope of breakthrough there is little useful prior knowledge which may be otherwise relied upon in determining this number of CFD.

In this paper, the question is analyzed in a very general setting and shown to be intractable. Next, a heuristic method is proposed in Sect. 2 as a first attempt to approximately solve the problem; and experimental results on selected datasets are presented to demonstrate the existence of a CFD for most of them. Coincidentally, it is observed that, for certain problems, only the CFD matters—in other words, random feature selection will be sufficient for good performance in data mining tasks provided that the number of features selected meets the CFD. In Sect. 3 the critical sampling problem is presented as one having exactly the same complexity as the CFD problem; and several heuristic methods for critical sampling are proposed, with experimental results showing that they are sufficient for practical purposes. Section 4 presents a set of data quality metrics for measuring the value of datasets from data mining perspectives.

2   The Critical Feature Problem

The feature selection problem has been studied extensively; and feature selection to satisfy certain optimal conditions has been proved to be NP-hard [5]. Here we consider the problem from a different perspective by asking the question whether there *exists* a CFD, i.e., a minimum number of features, that must be included for a data analytic task to achieve "satisfactory" results (e.g., building a learning machine classifier to achieve a given accuracy threshold), and we show the problem is intractable as it is in fact both NP-hard and coNP-hard.

Assume the dataset is represented as the typical n by p matrix $D_{n,p}$ with n objects (or data points, vectors, etc.) and p features (or attributes, etc.). The intuitive concept of the CFD of a dataset with p features is that there may exist, with respect to a specific "machine" M and a fixed performance threshold T, a unique number $\mu \leq p$ such that the performance of M exceeds T when a suitable set of μ features is selected and used (and the rest $p - \mu$ features discarded); further, the performance of M is always below T when any feature set with less than μ features is used. Thus, μ is the critical (or absolute minimal) number of features that are necessary to ensure that the performance of M meets the given threshold T.

Formally, for dataset D_p with p features (the number of objects in the dataset, n, is considered fixed here and therefore dropped as a subscript of the data matrix $D_{n,p}$), a machine M (a learning machine, a classifier, an algorithm, etc.) and performance threshold T (the classification accuracy of M, etc.), we call μ (an integer between 1 and p) the *T-Critical Feature* Dimension of (D_p, M) if the following two conditions hold:

1. There exists D_μ, a μ-dimensional projection of D_p (i.e., D_μ contains μ of the p features) which lets M to achieve a performance of at least T, i.e., $(\exists D_\mu \subset D_p)[P_M(D_\mu) \geq T]$, where $P_M(D_\mu)$ denotes the performance of M on input dataset D_μ.
2. For all $j < \mu$, a j-dimensional projection of D_p fails to let M achieve performance of at least T, i.e., $(\forall D_j \subset D_p)[j < \mu \Rightarrow P_M(D_j) < T]$

To determine whether a CFD exists for a D_p and M combination is a very difficult problem. It is shown below that the problem belongs to complexity class $D^P = \{L_1 \cap L_2 | L_1 \in NP, L_2 \in coNP\}$ [7]. In fact, it is shown that the problem is D^P-hard.

Since NP and coNP are subclasses of D^P (note that D^P is not the same as NP $\cap$ coNP), the D^P-hardness of the CFD problem indicates that it is both NP-hard and coNP-hard, and likely to be intractable.

2.1   CFDP Is Hard

The Critical Feature Dimension Problem (CFDP) is stated formally as follows: Given a dataset D_p, a performance threshold T, an integer k $(1 < k \leq p)$, and a fixed machine M. Is k is the *T-critical feature dimension* of (D_p, M)?

The problem to decide if k is the *T-critical feature dimension* of the given dataset D_p belongs to the class D^P under the assumption that, given any $D_i \subset D_p$, whether $P_M(D_i) \geq T$ can be decided in polynomial (in p) time, i.e., the machine M can be trained and tested with D_i in polynomial time. Otherwise, the problem may belong to some larger class, e.g., Δ_2^p (see [4]). Note here that $(NP \cup coNP) \subseteq D^P \subseteq \Delta_2^p$ in the polynomial hierarchy of complexity classes.

To prove that the CFDP is a D^P-hard problem, we take a known D^P-complete problem and transform it into the CFDP. We begin by considering the maximal independent set problem: In an undirected graph, a Maximal Independent Set (MIS) is an independent set (see [4]) that is not a subset of any other independent set; a graph may have many MIS's.

EXACT-MIS Problem (EMIS) – Given a graph with n nodes, and $k \leq n$, decide if there is a MIS of size exactly k in the graph is a problem known to be

D^P-complete [7]. Due to space limitations, we only sketch how to transform the EMIS problem to the CFDP.

Given an instance of EMIS (a graph G with p nodes, and integer $k \leq p$), to construct the instance of the CFDP, let dataset D_p represent the given graph G with p nodes (e.g., D_p can be made to contain p data points, with p features, representing the symmetric adjacency matrix of G), let T be the value "T" from the binary range {T, F}, let $k = k$ be the value in the given instance of EMIS, and let M be an algorithm that decides if the dataset represents a MIS of size exactly k, if yes $P_M =$ "T", otherwise $P_M =$ "F", then a given instance of the D^P-complete EMIS problem is transformed into an instance of the CFDP.

Detailed examples that explain the proof can be found in [9]. The D^P-hardness of the CFDP indicates that it is both NP-hard and coNP-hard; therefore, it's most likely to be intractable (that is, unless P = NP).

2.2 Heuristic Solution for CFDP

From the analysis above it is clear that even deciding if a given number k is the CFD (for the given performance threshold T) is intractable; so, to determine what that number is for a dataset is certainly even more difficult. Nevertheless, a simple heuristic method is proposed in the following, which represents a practical approach in attempting to find the CFD of a given dataset and a given performance threshold with respect to a fixed learning machine.

Though the heuristic method described below can be seen as actually pertaining to a different definition of the CFD, we argue that it serves to validate the concept and we show that for most datasets with which experiments were conducted a CFD indeed exists. Finally, the μ determined by this heuristic method is hopefully close to the theoretically-defined CFD.

In the heuristic method, the CFD of a dataset is defined as that number (of features) where the performance of the learning machine would begin to drop notably below an acceptable threshold, and would not rise again to exceed the threshold. The features are initially sorted in descending order of significance and the feature set is reduced by deleting the least significant feature during each iteration of the experiment while performance of the machine is observed. (For cross validation purposes, therefore, multiple runs of experiments can be conducted: the same machine is used in conjunction with different feature ranking algorithms; and the same feature ranking algorithm is used in conjunction with different machines; then we can compare if different experiments resulted in similar values of the CFD—if so the notion that the dataset possesses a CFD becomes arguably more apparent.)

Critical Dimension Empirically Defined. Let $A = \{a_1, a_2, \ldots, a_p\}$ be the feature set where $a_1, a_2, \ldots, a_p$ are listed in order of decreasing importance as determined by some feature ranking algorithm R. Let $A_m = \{a_1, a_2, \ldots, a_m\}$, where $m \leq p$, be the set of m most important features. For a learning machine M and a feature ranking method R, we call μ ($\mu \leq p$) the *T-Critical Dimension* of (D_p, M) if the following conditions are satisfied: when M uses feature set A_μ the performance of M is T, and whenever M uses less than μ features its performance drops below T.

Learning and Ranking Algorithms. In the experiments the dataset is first classified by using six different algorithms, namely Bayes net, function, rule based, meta, lazy and decision tree learning machine algorithm. The machine with the best prediction accuracy is chosen as the classifier to find the CFD for that dataset.

2.3   Results

For the experiments reported below, the ranking algorithm is based on chi-squared statistics, which evaluates the worth of a feature by computing the value of the statistic with respect to the class. Note that in the heuristic method the performance threshold T will not be specified beforehand but will be determined during the iterative process where a learning machine classifier's performance is observed as the number of features is decreased. Three large datasets are used in the experiments, each is divided into 60% for training and 40% for testing. Six different models are built and retrained to get the best accuracy. The model that achieves the best accuracy is used to find the CFD.

Amazon 10,000 Dataset. The Amazon commerce reviews dataset is a write print dataset useful for purposes such as authorship identification of online texts, etc. Experiments were conducted to identify fifty authors in the dataset of online reviews. For each author 30 reviews were collected, totaling 1500. There are 10,000 attributes with numerical values for all features. This becomes a multiclass classification problem with 50 classes. The results are shown in Fig. 1, where a CFD is found at 2486 features. The justifications that this is the CFD are, firstly, from 2486 downward, the performance drops quickly and unlike the situation at around 9000 the performance never rises thereafter; secondly, the performance at feature size 2486 is only slightly lower than the highest observed performance (at around 9000 features). Another point at around 6000 may also be taken as the CFD; however, 2486 is deemed more "critical"? since there is a big difference between 6000 and 2486 but very little difference between the performances at these two points.

Amazon Ad or Non-Ad Dataset. The Amazon commerce reviews Internet advertisement dataset is a set of possible advertisements on web pages. The task is to predict whether an image is an advertisement ("ad") or not advertisement ("non-ad"). The dataset includes 459 ad and 2820 non-ad images. Only 3 of the 1558 attributes of the dataset are continuous values and the remaining are binary. It is also noteworthy that one or more of the three continuous-valued features are missing in 28% of the instances. The classification results are shown in Fig. 2 where a CFD at feature size 383 is seen.

Thrombin Dataset. The training set consists of 1909 compounds tested for their ability to bind to a target site on thrombin, a key receptor in blood clotting. Each compound is described by a feature vector containing a class value (A for active, I for inactive) and 139,351 binary features describing 3-dimensional properties of the molecule. The task is to determine which properties are critical and to learn to accurately predict the class value. The classification results are shown in Fig. 3, where a CFD of 8487 is apparent.

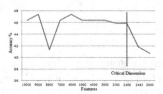

Fig. 1. The CFD of the Amazon 10,000 dataset.

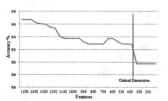

Fig. 2. The CFD of the Amazon Ad or Non-Ad dataset.

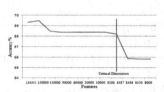

Fig. 3. The CFD of the Thrombin dataset.

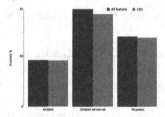

Fig. 4. Prediction accuracy at the CFD and at initial feature dimension (all features included).

We observed that each of the three datasets shows an apparent CFD, which is much smaller than the original feature dimension in each case while an acceptable level of performance is maintained. Figure 4 summarizes the results of experiments done. For additional reference, the results of 16 different datasets that were studied earlier can be found in [9].

Interestingly, it is observer that for certain data mining problems, if the dataset possesses a CFD of μ, then a random selected feature set with μ features will guarantee satisfactory performance in model building. In particular, the text classification problem is suspected to be such a problem where random feature selection may well be sufficient. Experiments are carried out on a set of 4 well-known corpuses of texts, using C4.5, KNN, and NB. Each time a different set of μ randomly selected features is used, and the performance is measured, the average of 9 experiments is considered the performance of the respective learning

Table 1. Results of random feature selection in text mining of four datasets.

Set	R8 (C4.5)	R8 (kNN)	WebKB	R52	News group
1	70.86	87.11	82.43	58.37	63.82
2	58.61	82.46	76.34	57.26	52.96
3	64.84	82.2	72.05	55.26	55.28
4	68.01	80.71	72.28	58.66	57.28
5	69.33	84.22	75.43	55.61	57.28
6	68.85	78.91	77.44	55.03	51.08
7	68.51	85.26	73.5	49.28	51.2
8	60.39	81.01	70.12	54.98	52.34
9	66.5	80.13	72.65	55.78	57.28
10	58.26	80.46	72.65	54.75	51.94
Ave	65.42	81.75	74.49	55.5	55.05

machine for the dataset. The results are summarized in Table 1 where row 1 lists results of using the top μ features, rows 2–10 are 9 experiments using randomly selected μ features, and the last row is the average of 9 experiments.

3   The Critical Sampling Problem

In this section, we consider the other problem of how to select a minimal sample of data points that will guarantee good performance. Other authors have addressed this problem [8]. Assume again the dataset is represented as an n by p matrix $D_{n,p}$. The concept of the Critical Sampling Size (CSS) of a dataset with n points is that there may exist, with respect to a specific machine M and a given performance threshold T, a unique number $v \leq n$ such that the performance of M exceeds T when some suitable sample of v data points is used; further, the performance of M is always below T when any sample with less than v data points is used. Thus, v is the critical (or absolute minimal) number of data points required in any sample to ensure that the performance of M meets the given threshold T.

Formally, for dataset D_n with n points (the number of features in the dataset, p, is considered fixed here when the only concern is the sample size, and therefore dropped as a subscript of the data matrix $D_{n,p}$). v, an integer between 1 and n, is called the *T-Critical Sampling Size* of (D_n, M) if the following two conditions hold:

1. There exists D_v, a v-point sampling of D_n (i.e., D_v contains v of the n vectors in D_n) which lets M to achieve a performance of at least T, i.e., $(\exists D_v \subset D_n)$ $[P_M(D_v) \geq T]$, where $P_M(D_v)$ denotes the performance of M on dataset D_v.
2. For all $j < v$; a j-point sampling of D_n fails to let M achieve performance of at least T, i.e., $(\forall D_j \subset D_n)$ $[j < v \Rightarrow P_M(D_j) < T]$

In the above, the specific meaning of $P_M(D_v)$, the performance of machine (or algorithm) M on sample D_v, is left to be defined by the user to reflect a consistent setup of the data analytic (e.g. data mining) task and the associated performance measure. The value of threshold T, which is to be specified by the user as well, represents a reasonable performance requirement or expectation of the specific data analytic task.

To determine whether a CSS exists, for a D_n and M combination, is a very difficult problem. Precisely, the problem of deciding, given D_n, T, k $(1 < k \leq n)$, and a fixed M, whether k is the T-critical sampling size of (D_n, M) belongs to the class D^P as well. In fact, it can be shown to be D^P-hard, exactly as the critical feature dimension problem (CFDP) analyzed in Sect. 2, and by using the same proof and merely selecting rows (instead of columns) of the adjacency matrix of the graph to construct a MIS.

3.1   Experimental Setup

Due to the complete symmetry or similarity to the CFD problem, it is suspected that simple heuristic methods can be developed to be sufficiently useful for practical purposes in solving the CSS problem, in the same way heuristic methods proved useful for finding the critical features. This section presents the experiments with a heuristic method on a few datasets downloaded from the UCI Machine Learning Repository [3]. Their dimensions vary both in the number of samples and in the number of features (descriptions of datasets can easily be found at the repository). This way, we are able to visualize the quality of the heuristic for different datasets.

Table 2. Datasets characteristics.

	Amazon ad non-ad	Credit	Hapt	Isolet
# Features	1558	23	561	617
# Classes	2	2	12	26
# Examples	3279	30000	10929	7797

3.2   Sampling Method

The size of both the training and test sets can influence the performance achieved by the learning machine. As a consequence when comparing the performance of a data mining task using the whole dataset and the sampled data, this must be taken in consideration. Because of this, three ways to split the datasets were considered, and therefore, for each of these methods, a respective CSS will be heuristically determined. Each method consists in splitting the dataset in different ratios: (30% Train/70% Test); (50% Train/50% Test); (70% Train/30% Test). In addition to ending up with a distinct CSS for each splitting ratio, the research design will also allow to observe their influence on the results.

The CSS is obtained iteratively. First the data is clustered using k-means and then, from each cluster, m examples are selected to form the sample D_v. In addition, regarding the way as the examples are sampled from each cluster, 3 orders were studied, (Asc) by ascending order of distance to the cluster centroid, ($Decr$) by decreasing order, and ($Rand$) randomly. To analyze the usefulness of this method, three approaches were used.

- $mk + r$: Initial approach, where the sample D_v is composed by selecting m examples from each one of the k clusters and then complemented with more $d * r$ random examples.
- mk: Same as above except that the sample D_v is not complemented with the extra random examples as the first method.
- r: Random sampling. The construction of D_v is made by randomly selected examples. In order to maintain some consistency, here r can be calculated with $m * k$.

3.3   Parameters Setup

The proposed heuristic to find the CSS has four parameters. Their values should be decided by considering (i) the nature of the problem, (ii) the size of the dataset, (iii) the data mining task that will be applied and (iv) the amount of available resources. These four parameters are, respectively, k, m, d and T. Next, we present some aspects to consider about these parameters.

Table 3. Settings for the heuristic parameters

Parameter	k	m	d	T
Description	# of clusters	# of instances from each cluster	# of instances ($*k$) to supplement sample	threshold value
Used value	2, 5, 10, 20, 30, 50	1% of cluster	$\frac{m}{4}$	$P_M(D_n)$

For larger datasets the values of k may increase. Both m and d must assume small values to allow us to monitor the evolution of performance of the critical sample as it grows at each iteration of the heuristic. To define T, $P_M(D_n)$ is the performance achieved by the learning machine using the whole dataset, for instance using 70% for training and 30% for testing. Regarding all these parameters, other values can and must be tested. As stated above, their values should be decided by considering various aspects.

3.4   Results

In Fig. 5, each dot of the lines represents the averaged CSS value, of 30 runs, for each value of k. For the $mk + r$ and mk sampling methods, the value that is represented is the one that obtained the best results among the *Asc, Decr* and *Rand* way to select the examples of the clusters. A total of 6 learning machines were tested, as can be seen in Table 4. The ones that obtained the best results were used. For binary datasets we used AdaBoost and for multiclass datasets, Multi Layer Perceptron. The performance were measured using F-score. Due to space limitations, in Fig. 5, only the CSS results, obtained when testing with 70% of the datasets, are shown. Despite this, the results for the remaining ratios were very similar.

Table 4. Learning algorithms used in the research design.

Algorithm	Learning machine
Lazy	KNN
Meta	AdaBoost
Neural Network	MLP
Tree	Decision tree random forest
Bayes	Naive Bayes

The results showed that the number of clusters, k, has a significant impact on the CSS, mainly on the Ads and Hapt datasets. It is possible to see the improvement of $mk + r$ and mk sampling methods compared to the random sampling, the r method. However, this is not so visible for the Ads and Isolet dataset. In the latter case, this can be due to the fact that the dataset is perfectly balanced, that is, each class has the same number of examples, which allows the random sampling to obtain good results. The Hapt dataset is also well balanced, however, it has a higher number of examples per class. Random sampling seems to be a good choice when dealing with datasets that contains low examples per class. The best method to select the examples from the clusters was the *Rand*.

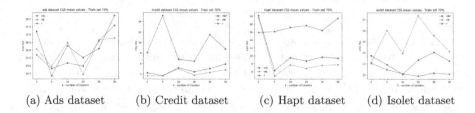

(a) Ads dataset   (b) Credit dataset   (c) Hapt dataset   (d) Isolet dataset

Fig. 5. Results of the 4 datasets with testing set of 70% of the data.

As the heuristic starts by having very small training sets, D_v, the training process is fast. Using this incremental approach, it is easier and faster to find out the CSS of a specific dataset. And by looking at these results, $mk + r$ and

mk sampling methods could be a good alternative to the random sampling. As for those two sampling methods, the best shows up to be mk since it presents more consistent results in the first 3 datasets. With these promising results, it is expected that good results will also be obtained for larger datasets.

Despite all this, the reduction for all datasets is noticeable. This shows that most of the data present in these datasets may be redundant, or even irrelevant.

4   Dataset Quality Metrics

As more and more networked physical objects become components of the Internet of Things and techniques of big data analytics advance (see [6]), the society is experiencing the transformation into a "data economy" where individuals, businesses, and organizations alike are contributing to both the generation and consumption of the big data.

There are various methods and techniques of data mining for different purposes; an ultimate goal of data analytics, however, is knowledge discovery for problems that have thus far defied solutions through conventional approaches.

In view of the above, a quantitative measure for the quality of datasets would be very useful for all concerned in big data analytics.

With regard to the capacity or potential for knowledge discovery or knowledge extraction from a dataset, we propose the following quality metrics for a given dataset $D_{n,p}$ (the dataset is represented as a matrix with n points, each represented as a p-dimensional vector).

From the concept of the critical sampling size discussed in Sect. 3, the *Sample Quality* of $D_{n,p}$ is defined as $Q_s = v/n$ where v is the critical sampling size of $D_{n,p}$. So, $v \leq n$; and $v = 0$, by definition, if the critical sample does not exist (or, equivalently, heuristic methods for finding v fail), in which case $Q_s = 0$ as well. In the optimal case, $v = n$ and so $Q_s = 1$, indicating that all data in the dataset are essential for the data mining task for knowledge discovery.

Likewise, the Feature Quality of $D_{n,p}$ is defined as $Q_f = \mu/p$ where μ is the critical feature dimension of $D_{n,p}$ (see [10]). So, $\mu \leq p$; and $\mu = 0$, by definition, if no critical feature sets exist, in which case $Q_f = 0$ as well, indicating that the dataset is insufficient due to the lack of certain features that are necessary. In the optimal case, $Q_f = 1$ when $\mu = p$, indicating that all the features are essential for the data mining for knowledge discovery task.

Q_D, the Overall Quality of the dataset $D_{n,p}$ (with respect to a specific learning machine M that is applied in mining it for model building and knowledge discovery), can therefore be defined as $Q_D = Q_s * Q_f = (v * \mu)/(n * p)$.

Note that $0 \leq Q_D \leq 1$. $Q_D = 0$ when either the critical feature set or critical sample does not exist (or cannot be found experimentally, with the respective heuristic methods employed to find them), indicating that the dataset is inadequate for the purpose of model building to achieve acceptable performance. At the other extreme, $Q_D = 1$ when $v = n$ and $\mu = p$, indicating that the dataset $D_{n,p}$ is indeed optimal, in terms of both the number of features and the number of data points, when it is evaluated with respect to the data mining task of model building and knowledge discovery for the problem under study.

5   Conclusions

This paper addresses the problem of finding a set of critical features and finding a critical sampling in data mining. It is shown that the two problems are highly similar in that both have the same complexity and theoretically intractable, yet both seem to allow simple heuristic methods for efficient and satisfactory solutions.

In view of the reported preliminary results, it is believed that the ongoing research on heuristic methods for determining critical feature dimensions and for finding critical sampling may lead to the development of effective solutions to cope with some of the challenges inherent in big data analytic problems due to the large dimensions of feature sets and the large number of samples in the big data. It is also hoped that the proposed metrics for data quality will be useful in areas like the "data on demand" service provided for data mining, model building or knowledge discovery purposes.

Acknowledgements. The authors gratefully acknowledge the reviewer of earlier versions of our papers whose insightful comments point them to fruitful directions of study, and their other colleagues and students who contributed to many helpful discussions and experiments.

References

1. Blum, A.L., Langley, P.: Selection of relevant features and examples in machine learning. Artif. Intell. **97**(1), 245–271 (1997)
2. Domingo, C., Gavaldà, R., Watanabe, O.: Adaptive sampling methods for scaling up knowledge discovery algorithms. Data Min. Knowl. Discov. **6**(2), 131–152 (2002)
3. UCI machine learning repository. http://archive.ics.uci.edu/ml. Accessed 20 Oct 2017
4. Garey, M.R., Johnson, D.S.: A Guide to the Theory of NP-Completeness, p. 70. WH Freemann, New York (1979)
5. Guyon, I., Elisseeff, A.: An introduction to variable and feature selection. J. Mach. Learn. Res. **3**, 1157–1182 (2003)
6. National Research Council: Frontiers in massive data analysis. The National Academies Press, Washington, DC (2013)
7. Papadimitriou, C.H., Yannakakis, M.: The complexity of facets (and some facets of complexity). J. Comput. Syst. Sci. **28**, 244–259 (1984)
8. Provost, F., Jensen, D., Oates, T.: Progressive sampling. In: Liu, H., Motoda, H. (eds.) Instance Selection and Construction for Data Mining, pp. 151–170. Springer, Boston (2001)
9. Suryakumar, D.: The critical dimension problem; no compromise feature selection. Ph.D. dissertation. New Mexico Institute of Mining and Technology (2013)
10. Suryakumar, Divya, Sung, A. H., and Liu, Q.: The critical dimension problem: No compromise feature selection. In: Proceedings of eKNOW 2014, the Sixth International Conference on Information, Process, and Knowledge Management, pp. 145–151. IARIA, Barcelona, Spain (2014)

From Recognition to Generation Using Deep Learning: A Case Study with Video Generation

Vineeth N. Balasubramanian[✉]

Indian Institute of Technology, Hyderabad,
Kandi, Sangareddy 502285, Telangana, India
vineethnb@iith.ac.in

Abstract. This paper proposes two network architectures to perform video generation from captions using Variational Autoencoders. We adopt a new perspective towards video generation where we use attention as well as allow the captions to be combined with the long-term and short-term dependencies between video frames and thus generate a video in an incremental manner. Our experiments demonstrate our network architectures' ability to distinguish between objects, actions and interactions in a video and combine them to generate videos for unseen captions. Our second network also exhibits the capability to perform spatio-temporal style transfer when asked to generate videos for a sequence of captions. We also show that the network's ability to learn a latent representation allows it generate videos in an unsupervised manner and perform other tasks such as action recognition.

Keywords: Deep learning · Generative models
Variational Autoencoders · Attention · Video understanding

1 Introduction and Motivation

Over the last few decades, the main objectives of machine learning, a popular branch of Artificial Intelligence (AI), have focused around recognition of objects (including people), their activities and interactions. Over the last few years, significant successes have been achieved in this area thanks to the emergence of deep learning, including object detection, object recognition and semantic segmentation. However, in the pursuit of AI to match human intelligence, recognition is but a small sub-goal, and higher forms of human intelligence remain a distant dream. One such form of human intelligence is the facet of creation or generation. While a human at a young age can recognize objects, their activities and interactions, it is non-trivial to generate the same entities. For instance, while it may be common to recognize an animal from a picture, it would be considered non-trivial for an average human to draw (or generate) a realistic image of an animal. A human would be considered talented (or of *'higher intelligence'*), if one possessed the ability to generate such images.

© Springer Nature Singapore Pte Ltd. 2018
G. Ganapathi et al. (Eds.): ICC3 2017, CCIS 844, pp. 25–36, 2018.
https://doi.org/10.1007/978-981-13-0716-4_3

While generative models have been a core component of machine learning methods over the last few decades, most methods such as Gaussian Mixture Models were used to model training data towards an objective such as classification or clustering, rather than being able to generate new data itself. The emergence of deep learning over the last few years has created a deep impact on this subgoal of AI too, through newer methods such as Generative Adversarial Networks (GANs) and Variational Autoencoders (VAEs). This work presents this perspective of deep learning, viz. its application to generating new data automatically. The generation of new data can be used for applying machine learning in application settings with low training data availability, as well as in domain adaptation settings where a different target domain has little training data available.

This work presents our recent work in the use of deep generative models for video generation, in particular, using VAEs. We present two methods, Sync-DRAW and ASImoV, towards this end. We show that the proposed methods generates natural-looking videos with the MNIST and KTH datasets. We also show how the proposed methods can automatically generate videos from text captions, as well as change the contents of the generated video in a seamless manner when the caption is changed midway during generation (called 'spatiotemporal style transfer' in our work).

The remainder of this article is organized as follows. A background of related methods for deep generative models is presented in Sect. 2; the proposed methods are presented in Sect. 3; the experimental results are presented in Sect. 4, and we conclude with pointers to future work in Sect. 5.

2 Background and Related Work

Recent years have witnessed the emergence of generative models like Variational Autoencoders (VAEs) [1], Generative Adversarial Networks (GANs) [2] and other methods such as using autoregression [3]. Earlier generative models based on methods such as Boltzmann Machines (RBMs) [4,5] and Deep Belief Nets [6] were always constrained by issues such as intractability of sampling. Recent methods have enabled learning a much more robust latent representation of the given data [7–10] and are helping improving the performance of several supervised learning tasks [11–13].

Previous approaches for image generation have extended VAEs and GANs to generate images based on captions by conditioning them with textual information [14,15]. Our approach also constitutes a variant of Conditional VAE [16] but differs in that in our paper, we use it to generate videos (earlier efforts only generated images) from captions with the capability to generate videos of arbitrary lengths. Further, our approach is distinct in that it incorporates captions by learning an attention-based embedding over them by leveraging long-term and short-term dependencies within the video to generate coherent videos based on captions.

The proposed methodology draws some similarity with past methods [17,18] that perform unsupervised video generation (but without captions), with all of them using GANs. Vondrick et al. [17] use a convolutional network with a fractional stride as the generator and a spatio-temporal convolutional network as a discriminator in a two-stream approach where the background and the foreground of a scene are processed separately. Saito et al. [18] introduced a network called Temporal GAN where they use a 1D deconvolutional network to output a set of latent variables, each corresponding to a separate frame and an image generator which translates them to a 2D frame of the video. However, in addition to not incorporating caption-based video generation, these approaches suffer from the drawback of not being scalable in generating arbitrarily long videos. Our approach, by approaching video generation frame-by-frame and utilizing the long-term and short-term context separately, effectively counters these limitations of earlier work. Besides, our approach learns to focus on separate objects/glimpses in a frame unlike [19] where the focus is on individual pixels, and [17] where network's attention is divided into just background and foreground.

In addition to generative modeling of videos, learning their underlying representations has been used to assist several supervised learning tasks such as future prediction and action recognition. [20] is one of the earliest efforts in learning deep representations for videos and utilizes Long Short-Term Memory units (LSTMs) [21] to predict future frames given a set of input frames. More recent efforts on video prediction have attempted to factorize the joint likelihood of the video to predict the future frames [19] or use the first image of the video with Conditional VAE to predict the motion trajectories and use them to generate subsequent frames [22]. We later demonstrate that in addition to the primary application of semantically generating videos with attentive captioning, it is possible to make small changes to our network which enable it to even predict future frames given an input video sequence.

3   Proposed Methods: SyncDRAW and ASImoV

3.1   Sync-DRAW

Let $X = \{x_1, x_2, \cdots, x_N\}$ be a video comprising of frames $x_i, i = 1, \cdots, N$ where N is the number of frames in the video. Let the dimensions for every frame be denoted by $A \times B$. We generate X over T timesteps, where at each timestep t, canvases for all frames are generated in synchronization. The Sync-DRAW architecture comprises of: (i) a *read mechanism* which takes a region of interest (RoI) from each frame of the video; (ii) the Recurrent VAE (*R-VAE*), which is responsible to learn a latent distribution for the videos from these RoIs; and (iii) a *write mechanism* which generates a canvas focusing on a RoI (which can be different from the reading RoI) for each frame.

During training, at each time step, a RoI from each frame of the original video is extracted using the read mechanism and is fed to the R-VAE to learn the latent distribution, Z, of the video data. Then, a random sample from this distribution, Z, is used to generate the frames of the video using the frame-wise

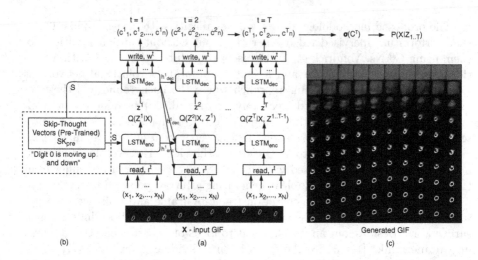

Fig. 1. Sync-DRAW architecture. (a) The read-write mechanisms and the R-VAE; (b) Conditioning the R-VAE's decoder on text features extracted from captions for text-to-video generation; (c) Canvases generated by Sync-DRAW over time (each row is a video at a particular time step, last row is the final generated video).

write mechanisms. At test time, the read mechanism and the encoder of the R-VAE are removed; and at each time step, a sample is drawn from the learned Z distribution to generate a new video through the write mechanism. We now describe each of these components. For details about the components including the read/write mechanisms, the R-VAE, how we integrate captions in the model as well as the loss function, please refer [23] (Fig. 1).

3.2   ASImoV

Let the random variable $Y = \{Y_1, Y_2, \cdots, Y_n\}$ denote the distribution over videos with $Y_1, Y_2, \cdots, Y_n$ being the ordered set of n frames comprising the videos. Let $X = \{X_1, X_2, \cdots, X_m\}$ be the random variable for the distribution over text captions with $X_1, X_2, \cdots, X_m$ being the ordered set of m words constituting the caption. $P(Y|X)$ then captures the conditional distribution of generating some video in Y for a given caption in X. Our objective is to maximize the likelihood of generating an appropriate video for a given caption. Since we would like to generate a video for a caption one frame at a time, we redefine $P(Y|X)$ as:

$$P(Y|X) = \prod_{i=1}^{n} P(Y_i|Y_{i-1}, \cdots\cdots, Y_1, X) \tag{1}$$

where n is the total number of frames in the video. Generation of the kth frame, Y_k, can therefore be expressed as:

$$P(Y_k|X) = P(Y_k|Y_{k-1}, \cdots, Y_1, X) \tag{2}$$

thus allowing the generation of a given frame to depend on all the previously generated frames. The generation can hence model both short-term and long-term spatio-temporal context. Y_k gathers short-term context, which consists of the local spatial and temporal correlations existing between any two consecutive frames, from Y_{k-1}. Y_k also obtains its long-term context from all the previous frames combined to understand the overall flow of the video. This ensures that the overall consistency of the video is maintained while generating the frame. In order to model this when generating the kth frame, we define two functions: U_k and V_k:

$$U_k = g(Y_{k-1}, X) \tag{3}$$
$$V_k = h(Y_{k-1}, \cdots, Y_1, X) \tag{4}$$

where U and V model the short-term and long-term stimulus respectively for generating Y_k. These two functions are implemented as new layers in our architecture, which is discussed in subsequent sections. Therefore, we now model $P(Y|X)$ as $P(Y|U,V) = \prod_{i=1}^{n} P(Y_i|U_i, V_i)$. For details about the components to implement the above formulation, including the R-VAE, the short-term and long-term attention mechanisms, and the loss function, please refer [24] (Fig. 2).

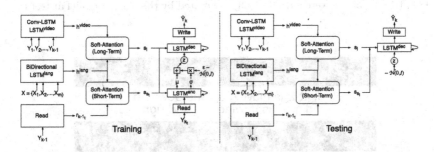

Fig. 2. Proposed network architecture for Attentive SemantIc Video (ASImoV) generation with captions. $Y = \{Y_1, Y_2, \cdots, Y_k\}$ denotes the video frames generated by the architecture, while $X = \{X_1, X_2, \cdots, X_m\}$ denotes the set of words in the caption.

4   Experiments and Results

4.1   Sync-DRAW Results

We studied the performance of Sync-DRAW on the following datasets with varying complexity: (i) Single-Digit Bouncing MNIST; (ii) Two-digit Bouncing MNIST; and (iii) KTH Human Action Dataset [25]. Considering this is the first work in text-to-video and there is no dataset yet for such a work, we chose these datasets since they provide different complexities, and also allow for adding captions[1].

---

[1] The codes, the captioned datasets and other relevant materials are available at https://github.com/Singularity42/Sync-DRAW.

We manually created text captions for each of these datasets (described later) to demonstrate Sync-DRAW's ability to generate videos from captions. We varied the number of timesteps T and K for read and write attention parameters across the experiments based on the size of the frame and complexity of the dataset (grayscale or RGB). Stochastic Gradient Descent with Adam [26] was used for training with initial learning rate as 10^{-3}, β_1 as 0.5 and β_2 as 0.999. Additionally, to avoid gradient from exploding, a threshold of 10.0 was used.

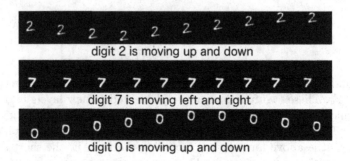

digit 2 is moving up and down

digit 7 is moving left and right

digit 0 is moving up and down

Fig. 3. Sync-DRAW generates videos from just captions on the Single-Digit Bouncing MNIST: Results above were automatically generated by the trained model at test time.

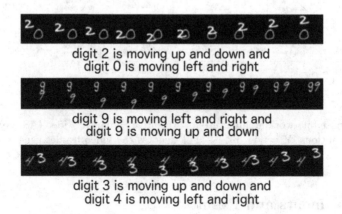

digit 2 is moving up and down and
digit 0 is moving left and right

digit 9 is moving left and right and
digit 9 is moving up and down

digit 3 is moving up and down and
digit 4 is moving left and right

Fig. 4. Sync-DRAW generates videos from just captions on the Two-Digit Bouncing MNIST: Results above were automatically generated by the trained model at test time.

Single-Digit Bouncing MNIST. For every video in the bouncing MNIST dataset, a sentence caption describing the video was included in the dataset. For Single-Digit Bouncing MNIST, the concomitant caption was of the form '*digit 0 is moving left and right*' or '*digit 9 is moving up and down*'. Hence, for the single-digit dataset, we have 20 different combinations of captions. In order

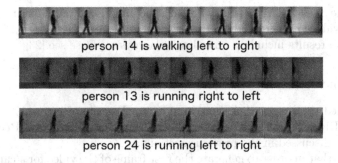

Fig. 5. Sync-DRAW generates videos from just captions on KTH dataset.

to challenge Sync-DRAW, we split our dataset in such a way that for all the digits, the training and the test sets contained different motions for the same digit, i.e. if a digit occurs with the motion involving *up and down* in the training set, the caption for the same digit with the motion *left and right* (which is not used for training) is used in the testing phase. Figure 3 shows some of the videos generated from captions present in the test set. It can be observed that even though the caption was not included in the training phase, Sync-DRAW is able to capture the implicit alignment between the caption, the digit and the movement fairly well.

Two-Digit Bouncing MNIST. We conducted similar experiments for the Two-Digit Bouncing MNIST, where the captions included the respective motion information of both the digits, for example '*digit 0 is moving up and down and digit 1 is moving left and right*'. Figure 4 shows the results on this dataset, which are fairly clear and good on visual inspection. These results give the indication that in the absence of captions (or additional stimuli), the attention mechanism in Sync-DRAW focuses on a small patch of a frame at a time, but possibly ignores the presence of different objects in the scene and visualizes the whole frame as one entity. However, by introducing captions, the attention mechanism receives the much needed "stimulus" to differentiate between the different objects and thereby cluster their generation, resulting in videos with better resolution. This is in concurrence with the very idea of an attention mechanism, which when guided by a stimulus, learns the spatio-temporal relationships in the video in a significantly better manner.

KTH Dataset. We were able to generate descriptive captions for the videos in the KTH dataset by using the metadata which included the person and the corresponding action. We carried out our experiments on videos for walking and running as it further helped to deterministically introduce the notion of direction. Some examples of the captions are '*person 1 is walking right to left*' and '*person 3 is running left to right*'. Figure 5 shows some of the videos generated by Sync-DRAW using just the captions for KTH dataset. The generated videos clearly demonstrate Sync-DRAW's ability to learn the underlying representation of even real-world videos and create high quality videos from text.

For more results including quantitative analysis, please see [23].

4.2 ASImoV Results

We evaluated the proposed ASImoV model on datasets of increasing complexity: Single-Digit MNIST, Two-Digit MNIST, and the KTH Action Recognition datasets, as discussed in the previous section[2].

We note that in order to generate the first frame of the video for a caption, we prefixed videos from all datasets with a start-of-video frame. This frame marks the beginning of every given video. It contains all $0s$ resembling the start-of-sentence tag used to identify the beginning of a sentence in Natural Language Processing [27].

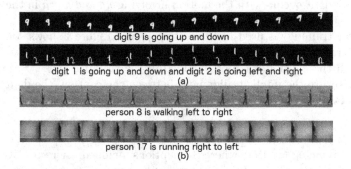

digit 9 is going up and down

digit 1 is going up and down and digit 2 is going left and right

(a)

person 8 is walking left to right

person 17 is running right to left

(b)

Fig. 6. The results of our network for different datasets to generate videos for a single caption. Number of frames generated is 15.

4.3 Results on Generation with Captions

The results on the Moving MNIST and KTH datasets are shown in Fig. 6, and illustrate the rather smooth generations of the model. (The results of how one frame of the video is generated over T time-steps is included in the supplementary material due to space constraints.) In order to test that the network is not memorizing, we split the captions into a training and test set. Therefore, if the training set contains the video having caption as 'digit 5 is moving up and down', the model at test-time is provided with 'digit 5 is moving left and right'. Similarly, in the KTH dataset, if the video pertaining to the caption 'person 1 is walking from left-to-right' belongs to the training set, the caption 'person 1 is walking right-to-left' is included at test time. We thus ensure that the network makes a video sequence from a caption that

---

[2] All the codes, videos and other resources are available at https://github.com/Singularity42/cap2vid.

it has never seen before. The results show that the network is indeed able to generate good quality videos for unseen captions. We infer that it is able to dissociate the motion information from the spatial information. In natural datasets where the background is also prominent, the network selectively attends to the object over the background information. The network achieves this without the need of externally separating the object information from the background information [17] or external motion trajectories [22]. Another observation is that the object consistency is maintained and the long-term fluidity of the motion is preserved. This shows that the long-term and the short-term context is effectively captured by the network architecture and they both work in 'coordination' to generate the frames. Also, as mentioned earlier, methodology ensures that a video can be generated with any number of frames. An example of a generation with variable length of frames is shown in Fig. 7.

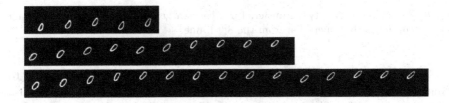

Fig. 7. Variable length video generation for caption 'digit 0 is going up and down'.

4.4   Results with Spatio-Temporal Style Transfer

We further evaluated the capability of the network architecture to generate videos where the captions are changed in the middle. Here, we propose two different settings: (1) *Action Transfer*, where the information of the motion that the object (i.e., digit/person) is performing is changed midway during the generation; and (2) *Object Transfer*, where the object information (i.e., digit/person) is changed midway during the generation. During action transfer, we expect that the object information remains intact but the motion that is being performed by the object changes; and during object transfer, we expect a change in the object information with respect to what has been generated so far. Results for action and object transfer can be seen in Figs. 8(a) and (b) respectively. We also go a step ahead and perform both *Action Transfer* and *Object Transfer* together as shown in Fig. 8(c). To test the robustness of the network, we ensured that the second caption used in this setup was not used for training the network.

We note here that when the spatio-temporal transfer happens, the object position remains the same in all the results, and the object from the second caption continues its action from exactly the same position. This is different from the case when the video is freshly generated using a caption since then the object can begin its motion from any arbitrary position. Moreover, the network maintains the context of the video while changing the object or action. For example,

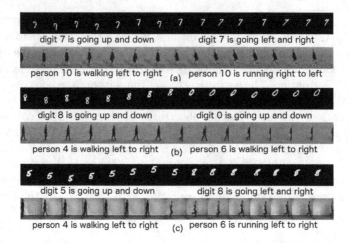

Fig. 8. Spatio-Temporal style transfer. First caption generates for 7 frames. Second caption continues the generation from the 8th frame.

in Fig. 8(c), the digit 5 with a certain stroke width and orientation changes to a digit 8 with the same stroke width and orientation. Similarly, in the natural dataset the type of the background and its illumination remains the same. The preservation of motion and context as well as the position is a crucial result in showcasing the ability of the network to maintain the long-term and short-term context in generating videos even when the caption is changed in the middle. For more results and quantitative results of this approach, please refer [24].

5   Conclusions and Future Work

In summary, we proposed network architectures that enable video generation from captions using VAEs, which is the first of its kind. Through various experiments, we conclude that our approaches, which combine the use of attention and short-term and long-term spatiotemporal context, are able to effectively to generate videos on unseen captions maintaining a strong consistency between consecutive frames. Moreover, the ASImoV network architecture is robust in transferring spatio-temporal style across frames when generating multiple frames. By learning using visual context, the network is even able to learn a robust latest representation over parts of videos, which is useful for tasks such as video prediction and action recognition. We did observe in our experiments that our network does exhibit some averaging effect when filling up the background occasionally. We infer that since the recurrent attention mechanism focuses primarily on objects, the approach may need to be extended with a mechanism to handle the background via a separate pathway. Future efforts in this direction will assist in bridging this gap as well as extending this work to address the broader challenge of learning with limited supervised data.

References

1. Kingma, D.P., Welling, M.: Auto-encoding variational bayes. arXiv preprint arXiv:1312.6114 (2013)
2. Goodfellow, I., Pouget-Abadie, J., Mirza, M., Xu, B., Warde-Farley, D., Ozair, S., Courville, A., Bengio, Y.: Generative adversarial nets. In: Advances in Neural Information Processing Systems, pp. 2672–2680 (2014)
3. van den Oord, A., Dieleman, S., Zen, H., Simonyan, K., Vinyals, O., Graves, A., Kalchbrenner, N., Senior, A., Kavukcuoglu, K.: Wavenet: a generative model for raw audio. CoRR abs/1609.03499 (2016)
4. Smolensky, P.: Information processing in dynamical systems: foundations of harmony theory. Technical report, DTIC Document (1986)
5. Salakhutdinov, R., Hinton, G.E.: Deep Boltzmann machines. In: AISTATS, vol. 1, p. 3 (2009)
6. Hinton, G.E., Osindero, S., Teh, Y.W.: A fast learning algorithm for deep belief nets. Neural Comput. 18(7), 1527–1554 (2006)
7. Radford, A., Metz, L., Chintala, S.: Unsupervised representation learning with deep convolutional generative adversarial networks. In: International Conference on Learning Representations (2015)
8. Radford, A., Metz, L., Chintala, S.: Unsupervised representation learning with deep convolutional generative adversarial networks. arXiv preprint arXiv:1511.06434 (2015)
9. Gregor, K., Danihelka, I., Graves, A., Rezende, D.J., Wierstra, D.: Draw: a recurrent neural network for image generation (2015)
10. Zhang, H., Xu, T., Li, H., Zhang, S., Huang, X., Wang, X., Metaxas, D.: Stackgan: text to photo-realistic image synthesis with stacked generative adversarial networks. arXiv preprint arXiv:1612.03242 (2016)
11. Ledig, C., Theis, L., Huszár, F., Caballero, J., Cunningham, A., Acosta, A., Aitken, A., Tejani, A., Totz, J., Wang, Z., et al.: Photo-realistic single image superresolution using a generative adversarial network. arXiv preprint arXiv:1609.04802 (2016)
12. Yeh, R., Chen, C., Lim, T.Y., Hasegawa-Johnson, M., Do, M.N.: Semantic image inpainting with perceptual and contextual losses. arXiv preprint arXiv:1607.07539 (2016)
13. Xu, K., Ba, J., Kiros, R., Cho, K., Courville, A., Salakhudinov, R., Zemel, R., Bengio, Y.: Show, attend and tell: neural image caption generation with visual attention. In: Proceedings of the 32nd International Conference on Machine Learning (ICML 2015), pp. 2048–2057 (2015)
14. Mansimov, E., Parisotto, E., Ba, J.L., Salakhutdinov, R.: Generating images from captions with attention. In: International Conference on Learning Representations (2016)
15. Reed, S., Akata, Z., Yan, X., Logeswaran, L., Schiele, B., Lee, H.: Generative adversarial text to image synthesis. In: Proceedings of the 33rd International Conference on Machine Learning, vol. 3 (2016)
16. Sohn, K., Lee, H., Yan, X.: Learning structured output representation using deep conditional generative models. In: Advances in Neural Information Processing Systems, pp. 3483–3491 (2015)
17. Vondrick, C., Pirsiavash, H., Torralba, A.: Generating videos with scene dynamics. In: Advances in Neural Information Processing Systems, pp. 613–621 (2016)

18. Saito, M., Matsumoto, E.: Temporal generative adversarial nets. arXiv preprint arXiv:1611.06624 (2016)
19. Kalchbrenner, N., van den Oord, A., Simonyan, K., Danihelka, I., Vinyals, O., Graves, A., Kavukcuoglu, K.: Video pixel networks. arXiv preprint arXiv:1610.00527 (2016)
20. Srivastava, N., Mansimov, E., Salakhutdinov, R.: Unsupervised learning of video representations using LSTMS. In: ICML, pp. 843–852 (2015)
21. Hochreiter, S., Schmidhuber, J.: Long short-term memory. Neural Comput. **9**(8), 1735–1780 (1997)
22. Walker, J., Doersch, C., Gupta, A., Hebert, M.: An uncertain future: forecasting from static images using variational autoencoders. In: Leibe, B., Matas, J., Sebe, N., Welling, M. (eds.) ECCV 2016. LNCS, vol. 9911, pp. 835–851. Springer, Cham (2016). https://doi.org/10.1007/978-3-319-46478-7_51
23. Mittal, G., Marwah, T., Balasubramanian, V.N.: Sync-draw: automatic video generation using deep recurrent attentive architectures. In: ACM Multimedia (2017)
24. Marwah, T., Mittal, G., Balasubramanian, V.N.: Attentive semantic video generation using captions. In: IEEE ICCV (2017)
25. Schuldt, C., Laptev, I., Caputo, B.: Recognizing human actions: a local SVM approach. In: Proceedings of the 17th International Conference on Pattern Recognition, ICPR 2004, vol. 3, pp. 32–36. IEEE (2004)
26. Kingma, D., Ba, J.: Adam: a method for stochastic optimization. In: International Conference on Learning Representations (ICLR) (2015)
27. Sutskever, I., Vinyals, O., Le, Q.V.: Sequence to sequence learning with neural networks. In: Advances in Neural Information Processing Systems, pp. 3104–3112 (2014)

Economic Dispatch Problem Using Clustered Firefly Algorithm for Wind Thermal Power System

K. Banumalar[(⊠)], B. V. Manikandan, and S. Sundara Mahalingam

Department of Electrical and Electronics Engineering,
Mepco Schlenk Engineering College, Sivakasi, Tamil Nadu, India
banumalar.234@gmail.com, bvmani73@yahoo.com,
radnus89@gmail.com

Abstract. In this paper, Clustered Firefly algorithm (CFFA) for the optimal dispatch of wind thermal power system (WTPS) has been presented. CFFA is used to solve the proposed economic dispatch problem for both conventional power system, and wind-thermal power system. The main objective is to minimization of Total cost includes the cost of energy provided by thermal generating units, wind turbine generators (WTG), and the cost of reserves provided by conventional thermal generators. The feasibility and effectiveness of the proposed method is verified by IEEE Six unit system and Forty unit system and the results are compared with existing methods available in the literatures.

Keywords: Clustered Firefly algorithm · Economic dispatch problem
Wind thermal power system

1 Introduction

The economic load dispatch problem is the problem of varying the generator's real and reactive power generation so as to meet the demands and losses, with the minimum fuel cost of power generation under some constrains. Along with Spinning reserve (SR) is one of the important ancillary services for secure and reliable operation of power system in the presence of unforeseen events such as, generation and/or line outages, sudden load changes, or both [1]. It is necessary to incorporate wind and thermal generating units in classical economic dispatch problem due to the increase in the use of renewable energy sources. The cost of power generation will be reduced due to the renewable energy resources.

Several papers have been published that address the use of conventional and intelligence algorithms to solving scheduling problem based on minimizing the total cost. Techniques like dynamic programming [2], mixed integer programming (MIP) [3] and artificial intelligence techniques like, genetic algorithm (GA) [4], simulated annealing (SA) [5], evolutionary programming (EP) [6], particle swarm optimization [7], ant colony optimization (PSO) [8] and Cuckoo search (CS) algorithm [9] are commonly used.

© Springer Nature Singapore Pte Ltd. 2018
G. Ganapathi et al. (Eds.): ICC3 2017, CCIS 844, pp. 37–46, 2018.
https://doi.org/10.1007/978-981-13-0716-4_4

This paper presents a new algorithm that merges the clustered behavior in the Firfly Algorithm and use it to solve the Economic Load dispatch problem subject to spinning reserves (SRs) constrain on six unit and forty unit systems.

2 Problem Formulation of EDP

2.1 Objective Function

Minimize Total Cost (TC):

$$\sum_{i=1}^{N_g} [FC_i(P_{gi}) + SRC_i(P_{sri})] + \sum_{j=1}^{N_w} [(FC_{wj}(P_{wj}) + SRC_{r.wj}(P_{wj} - P_{wj.av}) + FC_{p.wj}(P_{wj.av} - P_{wj})] \quad (1)$$

The first term in Eq. (1) is the fuel cost of thermal generators considering valve-point effect, and is given by

$$FC_i(P_{gi}) = a_i + b_i P_{gi} + c_i P_{gi}^2 + |ei \times sin(f_i(P_{gi}^{min} - P_{gi}))| \quad (2)$$

The second term is the SR cost of thermal generators, and is given by

$$SRC_i(P_{sri}) = x_i + y_i P_{gi} \quad (3)$$

The third term in the above objective function (1) is the direct cost paid to the wind farm owner for a scheduled wind power and is given by

$$FC_{wj}(P_{wj}) = d_j P_{wj} \quad (4)$$

The fourth term in Eq. (1) is the reserve requirement cost, is given by

$$SRC_{r.wj}(P_{wj} - P_{wj.av}) = K_{r.j}(P_{wj} - P_{wj.av}) = K_{r.j} \quad (5)$$

The fifth term is the penalty cost and is given by

$$FC_{p.wj}(P_{wj.av} - P_{wj}) = K_{p.j}(P_{wj.av} - P_{wj}) = K_{p.j} \quad (6)$$

The above problem is solved by considering the equality and inequality constraints.

2.2 System Constraints

Power Balance Constraints

For Study 1 the power balance condition for Thermal Generators is expressed as

$$\sum_{i=1}^{N_g} P_{gi} = P_d + P_l \quad (7)$$

For Study 2 the power balance condition for wind thermal generators is expressed as

$$\sum_{i=1}^{N_g} P_{gi} + \sum_{j=1}^{N_w} P_{wj} = P_d + P_l \tag{8}$$

Total Spinning Reserve Requirement Constraint

$$TSR_{req} = P_{g,l\,arg} + \sum_{j=1}^{N_w} (P_{wj} - P_{wj.av}) \tag{9}$$

This TSR_{req} will be provided by the online conventional thermal generators, and it is expressed as

$$\sum_{i=1}^{N_g} P_{sri} = TSR_{req} \tag{10}$$

2.3   Unit Constraints

Generation of Real Power Constraints
Each generator output power is restricted to minimum, maximum limits given as

$$\max\left[P_{gi}^{min}, P_{gi}^0 - R_{gi}^{down}\right] \leq P_{gi} \leq \min\left[P_{gi}^{max}, P_{gi}^0 - R_{gi}^{up}\right] \tag{11}$$

$$P_{wj}^{min} \leq P_{wj} \leq P_{wj}^f \tag{12}$$

Generator Spinning Reserve (SR) Constraints

$$0 \leq P_{sri} \leq \min(R_{gi}^{up}, P_{sri}^{max}) \tag{13}$$

$$P_{sri}^{max} = P_{gi}^{max} - P_{gi} \tag{14}$$

3   Wind Speed Calculation

Generally estimation of wind profile by means of Wei-bull probability density function (PDF). When compared to other methods, this probability distribution method has more accuracy in wind speed forecasting. To obtain the value of reserve and penalty costs, it is necessary to assume the PDF for the wind power output and load forecasts. In this work, the Weibull PDF is used for wind speed calculation and then, transformed to corresponding wind power distribution for use in the ED model, because of its simplicity.

4 Firefly Algorithm

Nature-inspired methodologies are among the most powerful algorithms for optimization problems. Firefly algorithm is a novel nature-inspired algorithm inspired by social behavior of fireflies. By idealizing some of the flashing characteristics of fireflies, firefly-inspired algorithm is presented by Yang [10].

4.1 Clustered Firefly Algorithm

The proposed CFFA has three workforce such as leader, the follower and the freelancer. The pseudo-codes are given below:

Step 1. Initialize the random population of fireflies.
Step 2. Evaluate the population on the given function.
Step 3. Sort the population on the fitness values.
Step 4. The first 10% of the population (size can vary according to function) is called the Leader.
Step 5. The next 80% of the population (size can vary according to function) is called the follower.
Step 6. The last 10 % of the population (size can vary according to the function) is called the freelancer.
Step 7. To each of the agent in the leader group a set of fireflies from follower group is allotted and they together make a single sub-population.
for i=1 to max no of iterations.
{
    for j=1 to max no of sub-population
    {
        Run CFFA for each of the sub group.
    }
Evaluate each particle in the freelancer group.
Find the best fitness values obtained among all the sub groups and the freelancer group. Again randomly initialize the entire freelancer population. (But see to it that the best particle is from the freelancer group, then that is not deleted in the next iteration.)
}
Step 8. The best population is thus best fitness values at the end of the iterations.

Now there is no form of communication between the subpopulations so they would always be searching independently (and exploring), rather than exploiting at the same place. The problem of exploitation is solved by the subgroups themselves. They exploit their local search spaces to get a minimum out of that region while since they do not communicate the exploration happens.

4.2 Implementation of CFFA for EDP

CFF algorithm is implemented to determine the power output of each generating unit for a specified demand at a particular time interval H_{in} order to minimize the total generation cost. In this algorithm, the brightness of a firefly is determined by the value of the objective function.

Step by Step Procedure for CFFA

Step 1: Specify generator cost coefficients, generation power limits for each unit, network data and load at the particular period. Initialize the control parameters of the CFF algorithm and maximum number of iteration ($Iter_{max}$).

Step 2: Generate the initial random population of R initial solution represented by real values (generator output) such as $R = [Y_1; Y_2; .....; Y_m]$ of m solutions in the multi dimensional solution space. Here m represents the firefly size.

Step 3: Modification of firefly position

For the EDP, Modify firefly position by Eq. (15)

$$P'_p = P_p + \beta(r) \times (P_p - P_f) + \alpha_f\left(rand - \frac{1}{2}\right) \tag{15}$$

Where $p, f \in \{1, 2 \ldots m\}$ are randomly chosen indexes. Although f is determined randomly, it has to be different from p.

Step 4: Repair Strategy

When firefly position is modified and these steps are given in Fig. 1.

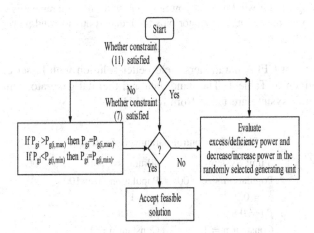

Fig. 1. Repair strategy for constraint management in EDP

Step 5: Evaluate the fitness of the population using Eq. (16). Here fitness of each population corresponds to the brightness of the firefly.

$$
\begin{aligned}
FIT_p &= 1/Cost_p, \qquad for\ Cost_p > 0 \\
&= 1 + abs(Cost_p), \quad for\ Cost_p < 0
\end{aligned}
\tag{16}
$$

$Cost_p$ is the cost of generation of pth string of the population. Determine the best fitness for individuals, and the corresponding minimum cost and the parameters responsible for the minimum cost.

Step 6: Memorize the best solution achieved so far. Increment the iteration.

Step 7: Stop the process if the termination criteria are satisfied. Termination criteria used in this work is the specified maximum number of iteration. Otherwise, go to step 3. The best fitness and the corresponding firefly position are retained in the memory at the end of the termination criteria is selected as the optimum power dispatched by the generating units when solving EDP for the specified time interval.

5 Results and Discussion

Four different test cases are considered for solving the ED problem for wind thermal power system using CFFA are considered below.

Case 1	Solving ED for thermal power system with SR requirement is equal to 10% of total demand. (without considering wind power)
Case 2	Solving ED for conventional power system with SR requirement is equal to the largest on-line thermal generator. (without considering wind power)
Case 3	Solving ED for wind-thermal power system with SR requirement is equal to 10% of total demand plus SR required due to wind power forecast uncertainty
Case 4	Solving ED for wind-thermal power system with SR requirement is equal to the largest on-line thermal generator plus SR required due to wind power forecast uncertainty

Proper setting of CFFA parameters yield better solution with lesser computational time and it is given in Table 1. The wind data and thermal generator unit data for six unit and forty unit system are taken from ref. [11]

Table 1. Control parameters for CFFA

6 unit System	40 unit system
Population size: 100	Population size: 100
$\gamma = 0.9$	$\gamma = 1$
$\beta 0 = 0.8$	$\beta 0 = 0.8$
Constant n = 1	Constant n = 1
No. of leaders: 10	No. of leaders: 10
No. of followers: 80	No. of followers: 80
No. of free lancers: 10	No. of free lancers: 10
No. of iterations: 500	No. of iterations: 500

5.1 Six Unit System

Table 2 shows the scheduled power generations, SRs and their cost for four different cases. The results obtained from CFFA are compared with existing method available in the literature are given in Table 3.

Table 2. ED solution for wind thermal power system

	Unit No.	P_G MW	$C(P_G)$ $	P_{SR} MW	$C(P_{SR})$ $		Unit No.	P_G MW	$C(P_G)$ $	P_{SR} MW	$C(P_{SR})$ $
Case 1	Pg1	151.51	921.90	13.34	30.46	Case 3	Pg1	161.8	1004.5	15.00	37.588
	Pg2	50.00	190	15.00	35.28		Pg2	43.22	192.37	14.48	36.465
	Pg5	31.27	154.89	0	25		Pg5	33.41	131.75	16.50	26.122
	Pg8	23.00	56.16	0	30		Pg8	22.64	57.346	7.963	30.84
	Pg11	18.00	44.1	0	25		Pg11	17.06	43.082	0	0
	Pg13	25.00	90.62	0	30		Pg13	19.07	61.991	0	0
	TC	**1633.41$**					**TC**	**1622.07$**			
Case 2	Pg1	151.67	922.18	14.95	30.01	Case 4	Pg1	161.8	1021.3	15.00	38.65
	Pg2	50.00	190.13	14.99	35.34		Pg2	43.22	170.81	14.48	36.3
	Pg5	31.09	154.84	10.83	25.55		Pg5	33.41	171.02	16.50	26.03
	Pg8	23.00	56.16	4.63	30.05		Pg8	22.64	55.74	7.96	30.49
	Pg11	18.00	44.08	0.51	25.08		Pg11	17.06	42.65	0	0
	Pg13	25.00	90.41	4.09	30.16		Pg13	19.07	56.56	0	0
	TC	**1633.99$**					**TC**	**1649.58$**			

Table 3. Comparison results of Six Unit System

Algorithm	Total cost ($)			
	Case 1	Case 2	Case 3	Case 4
PSO[11]	1642.73	1645.89	1646.42	1662.44
DE[11]	1641.86	1643.38	1645.55	1660.21
CMA-ES[11]	1637.97	1637.99	1641.97	1657.25
CMA-ES with MLT[11]	1633.45	1633.97	1636.57	1651.81
CFFA (Proposed Algorithm)	**1633.41**	**1633.99**	**1622.07**	**1649.58**

5.2 Forty Unit System

Table 4 shows the optimal total cost for four cases using CFFA and results are compared with other techniques available in the literature.

Table 4. Comparison results of 40 unit system

Algorithm	Total cost ($)			
	Case 1	Case 2	Case 3	Case 4
CMA-ES with MLT [11]	172256.74	165262.63	174197.78	161713.34
CFFA (Proposed Algorithm)	169814.02	148182.0	169781.00	161067.12

5.3 Computational Efficiency of CFFA

From Table 5, it is clear that CFFA is able to give a better solution with lesser computational time for six unit and forty unit system. The convergence graph of CFFA for Case 1 is shown in Figs. 2 and 3. From Tables 3 and 4, it is noted that significant amount of cost saving will be reflected per hour, per day and for per annum using proposed algorithm.

Table 5. Computational efficiency of CFFA

Test system	Mean time sec	Frequency of achieving the minimum cost better than the mean cost
6 unit system	12.31	16
40 unit system	22.10	18

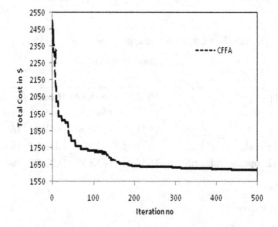

Fig. 2. Convergence graph - Six unit system

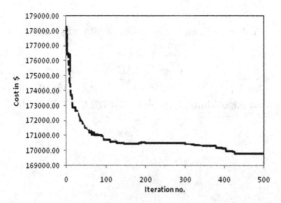

Fig. 3. Convergence graph - Forty unit system

6 Conclusion

The proposed clustered based Firefly algorithm is executed for ED problem with wind thermal power system and tested on six unit and forty unit system, successfully. In 6 unit system, least operating cost of $ 1622.07 is achieved. This leads to a social benefit of $ 14.5 when compared with CMA-ES with MLT. In forty unit system, a least operating cost of $169781.00 is achieved. This leads to a cost saving of $4416.78, when compared with CMA-ES with MLT, which is considered to be the best result available in the literature. From the results shows that economic dispatch problem containing wind power generation is economically significant, quite practical and perfect method in power system.

Abbreviation

N_g, N_w:	Number of Thermal and wind generator
P_{gi}:	Power generated by the unit i;
P_{Wj} and $P_{Wj.av}$:	Scheduled and available wind power from j^{th} wind power generator;
a_i, b_i, and c_i:	Cost coefficients of ith thermal generator;
e_i, f_i:	Coefficient of generators reflecting valve-point effect;
$k_{r.j}$ and $k_{p.j}$:	Reserve and Penalty cost coefficient for j^{th} wind generator;
P_d and P_l:	System demand and Transmission line losses (P_l) of power output
P_{Gi}^0:	Output power from i^{th} conventional generator in previous hour;
P_{Gi}^{min} and P_{Gi}^{max}:	Minimum and Maximum generation limits of thermal generators;
R_{Gi}^{up} and R_{Gi}^{down}:	Ramp up and ramp down limits of thermal generators (MW/h);
P_{wj}^{min} and P_{wj}^{f}:	Minimum and forecasted wind power from j^{th} wind generator;
P_{sri}^{max}:	Maximum reserve capacity.

References

1. Rebours, Y.A : Comprehensive assessment of market for frequency and voltage control ancillary services. Ph.D. thesis, University of Manchester (2008)
2. Balamurugan, R., Subramanian, S.: An improved dynamic programming approach to economic power dispatch with generator constraints and transmission losses. J. Electr. Eng. Technol. **3**(3), 320–330 (2008)
3. Lopez, J.A., Ceciliano-Meza, J.L., Moya, I.G., Gomez, R.N.: A MIQCP formulation to solve the unit commitment problem for largescale power systems. Int. J. Electr. Power Energy Syst. **36**(1), 68–75 (2012)
4. Abookazemi, K., Ahmad, H., Tavakolpour, A., Hassan, M.Y.: Unit commitment solution using an optimized genetic system. Int. J. Electr. Power Energy Syst. **33**(4), 969–975 (2011)
5. Rajan, C.C.A., Mohan, M.R.: An evolutionary programming based simulated annealing method for solving the unit commitment problem. Int. J. Electr. Power Energy Syst. **29**(7), 540–550 (2007)
6. Juste, K.A., Kita, H., Tanaka, E., Hasegawa, J.: An evolutionary programming to the unit commitment problem. IEEE Trans. Power Syst. **14**(4), 1452–1459 (1999)

7. Selvakumar, A.I., Thanushkodi, K.: A new particle swarm optimization solution to nonconvex economic dispatch problems. IEEE Trans. Power Syst. **22**(1), 42–51 (2007)
8. Simon, S.P., Padhy, N.P., Anand, R.S.: An ant colony system approach for unit commitment problem. Int. J. Electr. Power Energy Syst. **28**(5), 315–323 (2006)
9. Le, D.A., Vo, D.N.: Cuckoo search algorithm application for economic dispatch with wind farm in power systems. Global J. Technol. Optim. **7**, 1–6 (2016)
10. Yang, Xin-She: Firefly algorithms for multimodal optimization. In: Watanabe, Osamu, Zeugmann, Thomas (eds.) SAGA 2009. LNCS, vol. 5792, pp. 169–178. Springer, Heidelberg (2009). https://doi.org/10.1007/978-3-642-04944-6_14
11. Reddy, S.S., Panigrahi, B.K., Kundu, R., Mukherjee, R., Debchoudhury, S.: Energy and spinning reserve scheduling for a wind-thermal power system using CMA-ES with mean learning technique. Electr. Power Energy Syst. **153**, 113–122 (2013)

A Novel and Efficient Multi-hop Routing Algorithm for Manet

K. Kotteswari[1](✉) and A. Bharathi[2]

[1] Department of Computer Science and Engineering, Annai Mira College of
Engineering & Technology, Vellore, India
kottikarthikeyan17@gmail.com
[2] Department of Information Technology, Bannari Amman Institute of
Technology, Sathyamangalam, India
bharathia@bitsathy.ac.in

Abstract. Due to the development in the internet field, the wireless technology
gives rise to many new applications. Mobile Ad-hoc Network (MANET) is
optimistic features for research and development in the wireless network. Due to
the increasing popularity of mobile devices, the wireless ad-hoc network has
become more active and dynamic network for communication. It is essential to
have high-performance MANET comprised of mobile nodes to develop a critical application in wireless technology. Certain Quality of Service (QoS) factors
that influence the performance of routing in mobile nodes include energy efficacy, Communication bandwidth, network security, and resource utility etc.
A novel routing algorithm namely Better Performance Provisioning Algorithm
(BPPA) is proposed. BPPA integrate certain parameters such as distance, node
behavior, direction, mobility, and hop count to accomplish high- performance
MANET. BPPA algorithm contemplates in calculating the distance and the
direction of target node using Average Time-of-Arrival (ATOA) algorithm and
Direction Estimation algorithm (DES) respectively. The nodes are clustered
using distance and mobility factor found with several behavioral attributes of the
nodes. The cluster pattern, Node Capacity Factor (NCF), and hop count algorithm are combined to estimate the contextually effective route. The data packet
is routed using the above effective route successfully from source to destiny. The
routing Overheads, throughput, and Packet Delivery Ratio are appraised to
compare the performance with traditional algorithms like DSDV, AODV, and
DSR.

Keywords: BPPA · ATOA · DES · NCF · Performance evaluation

1 Introduction

Among various wireless technologies, Mobile Adhoc Network (MANET) is more
popular and active network in the communication era. MANET is dynamic, self-
computing, and framework-less network with mobile nodes that connect each other
over a wireless link. In MANET, any mobile node has the liberty to meander inside and
outside the network at any instant of time. The primary purpose of MANET is to
employ itself in a critical application such as military scenarios, emergency purpose,

© Springer Nature Singapore Pte Ltd. 2018
G. Ganapathi et al. (Eds.): ICC3 2017, CCIS 844, pp. 47–57, 2018.
https://doi.org/10.1007/978-981-13-0716-4_5

data networks, device networks and commercial sectors etc. [1]. High performance is required for MANET to handle the above critical application. The main factors that can accomplish high performance in mobile nodes are bandwidth, security, energy and resource.

The goal of this process is to upgrade the MANET routing performance for sending data from source to destiny in the above mentioned critical application without any data loss, delay, congestion, and overheads. This goal can be accomplished by using Better Performance Provision Algorithm (BPPA). BPPA optimize the routing performance of mobile nodes and also targeted in refining the QoS. The process of organizing the packet across the network from one host to another host is supervised by routing algorithm. In routing algorithm, the factors used to accomplish the goal differ according to their scenario but their goal is to deliver data packet or signal to the target without any error. Some of the factors considered in the routing algorithm to optimize performance are as follows. The efficiency of a route [2] is decided by three factors individually: the route expiration time, the node behavior, and the hop count. In [3] Queuing delay, Energy cost, and link stability are routing factor considered for better performance. In [4] the author designed a network using hop count factor. The path with less hop count shows more throughput gain and limited energy consumption. The QoS factor [5] interpreted for improving the performance in MANET are network traffic, neighbor node behavior, mobility, and data repetition. This QoS measure generates high route performance in terms of network delay, routing overhead, and mobility. In [6] the distance and hop count is appraised for refining performance. The route with minimum hops and minimum distance are computed and by manipulating this route, the consumption of energy is reduced. The positions of nodes [8] are regulated using the direction and distance factor to show improvement in performance. In [7] the author describes the impact of bandwidth. A node with less bandwidth leads to poor routing performance, less throughput, and traverses with less no of hops. A network of nodes with low bandwidth decreases the performance in communication. Generally mobile nodes have limited energy capacity. Throughout the transaction, energy must be persuasively established for better performance [9].

The limitation in MANET is Throughput and Delivery Ratio of packet decreases at some threshold value that may diminish the MANET performance [10]. To overcome this limitation and to exhibit high performance in critical application, a new routing algorithm BPPA is employed which associate various factors to improve the performance. The various factors are distance, delay, direction, node capacity and hop count. In [11], the ATOA, DES algorithm for calculating distance and direction is obtained. The distance is estimated using Average-Time-of-Arrival (ATOA). ATOA uses Round Trip Time (RTT) and signal rate to determine the distance between nodes. The Node Based Clustering (NBC) is used to cluster intermediate nodes into three NBC level, according to their mobility. Direction Estimation Algorithm (DES) is defined to compute the direction of the destiny using the distance between nodes. By estimating distance and direction of destiny, a part of network area where the target node exists is only considered and remaining part is eliminated. So that energy consumption and resource utilization can be limited. Higher the network density leads to the lower throughput of the network. To handle the network density and mobility, Node Capacity Factor (NCF) is introduced. NCF calculates the buffer capacity and behavior of nodes in the network. The buffer capacity is about the memory size of nodes which is used to handle network capacity and to avoid congestion in the network. The node's behavior is predicted from the history of previous transactions maintained in a table. The table is formed for every node that participates in the network. The appropriate node that are capable and effective for communication are selected based on NBC levels and NCF. The multiple routes are obtained from these appropriate nodes. An Efficient route is determined from the obtained multiple routes using hop count factor.

The benefits of proposed work is to (i) Integrate various factors that improves the network performance and influence routing strategy, (ii) Introduce clustering method to differentiate the parameters like node's behavior, memory capacity, mobility and distance between those nodes that participate in network communication, (iii) Filter the incapable nodes that consume more energy while transferring data packet through direction of destiny, (iv) Handles mobility and Network density using capacity and behavior of nodes, and (v) Determine the effective path which leads to better performance using hop count. The remaining section of this process is arranged as follows: Sect. 2 describes a review of the previously used method that is used to improve performance. Section 3 presents overview and details of the novel BPPA routing. Section 4 presents implementation of proposed BPPA algorithm. Section 5 analyzes the performance of MANET using the metric considered for it. Section 6 concludes with the impact of BPPA and the future enhancement of proposed work.

2   Related Work

Currently, many investigations have been undergone to acquaint various factors that have the ability to influence the routing performance in MANET. In each investigation, they use many factors separately to achieve their goals. A localization evolution technique was introduced by Zekavat et al. [14]. In many scenarios, Global Position System (GPS) is employed in determining the location of mobile nodes in MANET but here Direction-of-Arrival (DOA) and Time-of-arrival (TOA) is employed. A pertinent shortest path is determined for reaching destiny using the location of each node. In another approach, Ashraf et al. [18] present location evaluation approach based on trilateration technique. The location of mobile nodes is estimated using the received signal rate (RSS). Comparing to TOA the RSS shows less performance. Martín-Escalona et al. [15] examines the performance of WLAN with IEEE 802.11v based on Time-of-Arrival (TOA). TOA is determined by the Round-Trip-Time (RTT). RTT is a time required for a pulse or packet to travel from the specific source node to the destination node. The different routing issues are power utilization, neighbour node selection, and load stabilization. To conquer these problems, different factors are separately utilized for optimizing the routing performance. Based on congestion detection algorithm, Senthil Kumaran et al. [16] reduces the congestion in the network to improve routing performance. Congestion in the network causes packet loss, delay, and buffer overflow. The congestion can calculated from buffer size of nodes. From the same perspective Zhang et al. [17] has proposed a new Hybrid On-demand Distance Vector Multi-path (HODVM) routing protocol that uses the non-congestion degree to balance the network load. Non-congestion degree is established using the buffer capacity. The Route Request (RREQ) message and Route Reply (RREP) message gather information about buffer capacity of nodes.

Sexton et al. [13] utilizes new method related to mobility factor called Path Encounter Rate (PER). PER is evaluated by identifying the alteration of adjacent mobile node within a time period to determine the stability of wireless link before forwarding a data packet (or) signal. In [12], the factor considered is hop count. The path in which has less hop count is defined to have more performance in MANET. The shortest-path between source and destination is estimated using the hop count factor. In the overview of above routing algorithm, distance, direction, mobility, node capacity, and hop count are mainly considered as factors that are used to optimize performance. The advantage of proposed work is that it combines the factors that are used to influence the various QoS factors and can survive in any critical application without any issues.

3 An Overview of Better Performance Provisioning Algorithm

The various notation involved in BPPA is described as S and D indicate Source node and Destination node respectively. The intermediate node is indicated by $\{n_1, n_2 \ldots n_i, \ldots, n_k\}$. D_N, C_N, M_N and S_R indicate Distance between nodes, clustering of nodes, Mobility factor of nodes & Signal Rate. The angle between S and D is represented as Θ. TB_N, UB_N, NB_N and NCF represents Total Buffer-capacity of nodes, Utilized buffer-capacity of nodes, Node Behavior & Node Capacity. The path formation using nodes denoted as P_E.

3.1 Distance Estimation Using ATOA

The Average Time-of-Arrival (ATOA) algorithm is introduced to find the distance between the nodes. ATOA provide more accuracy than TOA in calculating distance. To provide more accuracy, ATOA consider delay while establishing distance and take an average value of TOA. The ATOA can be obtained as follows:

$$TOA = [(RTT - t)/2] - DL \tag{1}$$

- $t \rightarrow$ Processing time
- $DL \rightarrow$ Delay Factor

The Round Trip Time (RTT) is calculated by forwarding RREQ (Route Request signal) from the originator to all adjacent nodes that participate in the network. The RREQ received nodes forwards RREP (Route Reply signal) to the originator. The time taken for forwarding and delivering the route signal from the specific source to destiny is called as RTT. Initially, the processing time t is considered as a constant value for all nodes. The delay factor DL is measured based on nodes performance in the network.

To maintain accuracy, the TOA is evaluated for different values of RTT and average of TOA is taken by combining all value of TOA. The average of five times is based in network density

$$ATOA = (\text{Sum of TOA}) / 5 \tag{2}$$

The distance between sources to all the other nodes is obtained using Signal rate S_R and ATOA value.

$$\text{Distance, } D_N = S_R * ATOA \tag{3}$$

3.2 Node Based Clustering (NBC)

To the best of our knowledge, the clustering concept is very less used in the routing algorithm of MANET. The clustering is used in BPPA to classify the nodes, according to their speed or mobility of the nodes. This algorithm can handle the mobility in the routing algorithm. The mobility factor (M_N) is defined as the distance travelled within the duration of time. The nodes are clustered into three levels based on their mobility factor. The NBC levels are Low, Medium, and High. These levels value differ for each network. In BPPA, clustering is done to filter the nodes which have more speed. These nodes will lead to linkage failure and will not deliver the packet because of its mobility. To overcome the clustering is done. The mobility factor (M_N) is calculated as

$$M_N = \text{Distance} / \text{time} \tag{4}$$

Instead of using any pre-defined clustering algorithm for clustering of nodes, a simple formula is employed for clustering based on the nodes mobility factor.

$$\text{Clustering of node, } C_N = D_N / M_N \tag{5}$$

3.3 Direction Estimation Algorithm (DES)

DES is used to estimate the direction of the target node. Many reference nodes are fixed in the network area and distance between the reference node and destiny is estimated by ATOA. Based on the distance of source, reference, and destiny, the direction of the destiny is evaluated as:

$$\Theta = \sin^{-1}[(b) / y] \tag{6}$$

- b → distance between S and D
- y → distance between D and R

Angle of Assumption for estimating direction of destination node are as follows. For 0–90°, the direction is I quadrant (North-East), for 90–180°, the direction is II quadrant

(South-East), for 180–270°, the direction is III quadrant (South-West), for 270–360°, the direction is IV quadrant (North-West).

The value of angle decides the position of destiny. After estimating the direction, the nodes that are present towards the destiny direction is only used for communication. Remaining nodes are eliminated. Unnecessary consumption of energy can be avoided by only using the filtered nodes for forwarding packets.

3.4   Node Capacity Factor (NCF)

The nodes are selected based on their Node Capacity Factor (NCF). The node which has less congestion and more behaviour is considered as appropriate nodes. The congestion occurs in the network is reduced, delay can be reduced and mobility of the node can be handled using NCF. The Node Capacity Factor can be obtained as:

$$NCF = [(TB_N) / (UB_N)] - NB_N \tag{7}$$

- Total Buffer-size, $TB_N \rightarrow$ Total memory size of each node
- Utilized-buffer, $UB_N \rightarrow$ Remaining memory-size after utilization
- Behaviour of node, $NB_N \rightarrow$ Behaviour of 10 transaction occurred in the node

The behaviour of a node is a new method maintained in a table that describes the previous transaction of control or data packet and their behaviour process. The process behaviour is to check whether the process of packets is failure or success without any data loss and delay. The behaviour of the node is determined using the percentage of success rate SX_R and failure rate F_R of the packet in those 10 transactions. The SX_R is number of signal or packets transmitted successfully in those 10 transactions without any data loss and delay. The Percentage of Packet success is calculated as:

$$\% \text{ of Packet Success rate} = [(10 - SX_R)/10] \times 100 \tag{8}$$

Based on the percentage obtained from the success rate the node behaviour of each node is determined as follows. For 100 to 95% the NB_N value is 5–4, 95 to 85% the value is 4–3, 85 to 75% the value is 3–2, 75 to 65 the value is 2–1, % less than 65 the value is 1–0. A node behaviour rating is maintained from 0 to 5 for each percentage of packet success rate obtained.

3.5   Hop Count (HC)

The Hop count is number of hops passed through the route. The increase in hop count leads to decrease in throughput. The route with fewer hops is appraised as an efficacy route P_E.

4   Implementation of BPPA

Let R $\{n_1, n_2 \dots n_i, \dots ., n_k\}$ be intermediate nodes from the source S to the intended destiny D. NCF and RTT set is created for all mobile nodes. The routing table RT_i $\{RT_1, RT_2, RT_3 \dots RT_n\}$ that contain nodes address, RTT, Delay DL, and NCF for each node is constructed.

4.1   Initiating and Processing Route Request

The source S broadcast RREQ (Route Request signal) to adjacent nodes $\{n_1, n_2, \dots n_i \dots n_k\}$ with a time limit Time To Live (TTL). TTL is a time period in which route signal broadcast each other between the nodes. After the completion of the time period, the nodes will stop receiving the signal and starts discarding the signal. RTT value is initialized to zero.

$$S \rightarrow \text{all neighbour node} \{n_1, n_2, \dots, n_i \dots n_k\},$$
$$RREQ = \{RREQ^* + TTL + NCF + RTT\}$$

where RREQ consists of original route request signal, Node Capacity factor, Round Trip Time and Time to Live.

When nodes receive RREQ, NCF_{RT} and RTT is updated to their corresponding RT_i and TTL time start decreasing. If the receiving node is destiny D, it generates and send Route Reply signal (RREP) to the neighbour nodes. If it is intermediate node $\{n_1, n_2, \dots, n_i \dots n_k\}$, then it forward the RREQ to next one-hop neighbour nodes.

4.2   Initiating and Processing Route Reply

The intermediate node $\{n_1, n_2 \dots, n_i \dots n_k\}$ and destination node that received RREQ will generate the RREP signal

$$D, \{n_1, n_2, \dots, n_i \dots n_k\} S \rightarrow, RREP = \{RREP * + NCF + RTT + TTL\}$$

where RREP is original route reply signal, Node capacity factor, Round Trip Time and Time To Live.

The NCF_{RREP} is initialized at this stage, and RTT value is added at each hop node in the routing table RT_i. TTL time keeps on decreasing. The NCF is set to zero if it is generated by destination or it is set to NCF_{RREP} value that acquired by the current node if it is generated by intermediate node $\{n_1, n_2, \dots, n_i, \dots, n_k\}$. When the node receives RREP, it creates or updates the NCF_{RREP} and RTT in the RT_i for their corresponding route. In RT_i, the NCF_{RT} and NCF_{RREP} of a node are compared to determine whether to update the larger value of NCF. The NCF value is updated in RT_i when it has larger value. If the receiving node is source node S, then the distance and direction is estimated by ATOA and DES respectively. If the receiving node is transitional node $\{n_1, n_2, \dots, n_i \dots n_k\}$, then the RREP message forwarded to next hop node towards source node S in the route and the NFC and RTT value is updated in RT_i. If the TTL value is zero, the originator node S stops receiving RREP signal and discard the signal. Delay factor DL is also estimated for each node and updated in the routing table RT_i (Fig. 1).

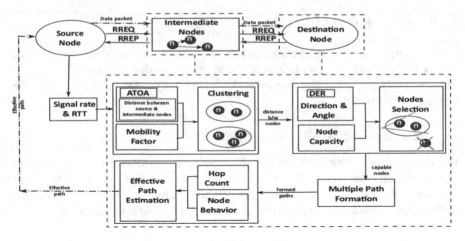

Fig. 1. Architecture of BPPA protocol

4.3 Exquisite of Nodes

From RT_i, the signal rate and Round Trip Time (RTT) are extracted from each node. Average-Time-of-Arrival (ATOA) algorithm is employed to establish distance between nodes by using RTT and DL. The nodes are classified into cluster based on Node Based Clustering (NBC). The direction is determined using the Direction Estimation Algorithm (DES). Only the nodes which have Medium level in NBC and present in the direction of target node are taken for path deployment. Multiple paths $\{P_1, P_2, P_3 \ldots P_N\}$ are developed.

4.4 Effective Path Formation and Data Packet Forwarding

After forming multiple paths using the suitable nodes, now it's time to find out the effective among the formed multiple paths. The Hop count factor is used to determine the effective route. In the multiple paths, each path that has less hop count is chosen. The link stability is also considered in this algorithm. The hop count algorithm is established to reduce packet drop and to provide more throughput. The route path is said to be efficient when it has less hop count and large value of NCF in P_i. Now the node S sends the data packet to intended destination D through this efficient route path P_E.

5 Simulation

In this portion, the simulation is done to evaluate the performance of Proposed Better Performance Provisioning Algorithm (BPPA). To evaluate the performance of BPPA, it is compared with other traditional routing like Dynamic Source Routing (DSR), Destination-Sequence Distance Vector Routing (DSDV) and Ad Hoc On-demand Distance Vector Routing (AODV) protocol. The Network Simulator (NS-2) is a network tool used for implementation.

5.1 Simulation Environment

The network area consists of 100 to 250 nodes in a 1900 × 800 m size. The radio range is 50 m with 2 Mbps bandwidth. The MAC layer is based on IEEE 802_11. The radio propagation model, TwoRayGround is used. The interface queue type used at MAC layer is PriQueue/DropTail that can hold 50 packets before they send to the physical layer. A routing buffer can store up to 64 packets in the network layer. An omnidirectional antenna is used to transfer data packet and signal. The routing protocols used are BPPA, AODV, DSR, and DSDV. The maximum speed of mobile nodes in the network is 10 to 50 m/s.

5.2 Performance Evaluation

The performance is estimated in terms of following metrics:

Average End-to-End Throughput (AET). The ratio of total number of packet (or) signal that reaches node D from the node S and total duration for node D to receive the last packet (or) signal.

$$AET = \frac{\sum \text{Number of packet or signal received}}{\text{Total duration for receiving packet or signal}}$$

Packet Delivery Ratio (PDR). The ratio of number of the packet (or) signal transferred by the node S and number of the packet (or) signal received by the node D.

$$PDR = \frac{\sum \text{Number of packet or signal delivered}}{\sum \text{Number of packet or signal sent}}$$

5.3 Simulation Result

Packet Delivery Ratio with Number of Nodes. Figure 2, Shows the PDR values with number of nodes. The PDR shows the accuracy of the packet in the network. The PDR percentage value of BPPA, AODV, DSR, and DSDV is 77, 67, 59, and 58. The PDR rate of BPPA is greater compared to AODV, DSR, and DSDV.

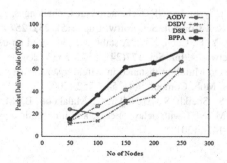

Fig. 2. Comparison of PDR with no of nodes

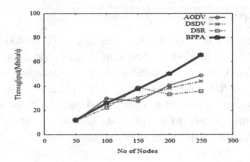

Fig. 3. Comparison of AET with no of nodes

Average End-to-end Throughput (AET) with Number of Nodes. The Average End-to-end Throughput (AET) is evaluated with varying number of nodes. The throughput directly depends on the performance. The AET rate of BPPA, AODV, DSR, and DSDV routing protocol is 66, 56, 48, and 44. The AET rate of BPPA is more compared to AODV, DSR, and DSDV as shown in Fig. 3.

6 Conclusions

This work introduces a new metric in MANET, Node Capacity Factor (NCF), Node Based Clustering, distance, direction and combine with Hop Count Factor (HC) that has ability to handle the mobility circumstance, perform well in critical application and network density. Thus, the performance of new routing algorithm Better Performance Provisioning Algorithm (BPPA) deployed with collective factors is compared with routing protocol constructed with Hop count metric. The analysis and simulation result shows that BPPA routing algorithm shows better performance. The NCF consists of node behavior and node buffer size. The node behavior deals with mobility. The DES reduces energy consumption. The NBC can handle the mobility of nodes. The Future enhancement of this work is dealing -with the security factor of the node which is not concentrated in this work. The data security can lead to better performance in MANET.

References

1. Aarti, Tyagi, S.S.: Study of MANET: characteristics, challenges, application and security attacks. Int. J. Adv. Res. Comput. Sci. Softw. Eng. **3**(5), 252–257 (2013)
2. Wang, N.-C., Huang, Y.-F., Chen, J.-C.: A stable weight-based on-demand routing protocol for mobile ad hoc networks. Inf. Sci. **177**(24), 5522–5537 (2007)
3. Vural, S., Ekici, E.: On Multihop distance in wireless sensor networks with random node location. IEEE Trans. Mob. Comput. **9**(4), 540–552 (2009)
4. Guo, Z., Malakooti, S., Sheikh, S., Al-Najjar, C., Malakooti, B.: Multi-objective OLSR for proactive routing in MANET with delay, energy, and link lifetime predictions. Appl. Math. Model. **35**(3), 1413–1426 (2011)

5. Zhang, Z., Mao, G., Anderson, B.D.O.: On the Hop count statistics in wireless multihop networks subject to fading. IEEE Trans. Parallel Distrib. Syst. **23**(7), 1275–1287 (2012)
6. Li, Z., Shen, H.: A QoS-oriented distributed routing protocol for hybrid wireless networks. IEEE Trans. Mob. Comput. **13**(3), 693–708 (2014)
7. Mukherjee, S., Avidor, D.: Connectivity and transmit-energy considerations between any pair of nodes in a wireless Adhoc network subject to fading. IEEE Trans. Veh. Tech. **57**(2), 1226–1242 (2008)
8. Vir, D., Agarwal, S.K., Imam, S.A.: Performance analysis of MANET with low bandwidth estimation. Int. J. Sci. Res. Pub. **3**(3), 1–5 (2013)
9. Kunz, T., Alhalimi, R.: Energy-efficient proactive routing in MANET: energy metrics accuracy. AdHoc Netw. **8**(7), 755–766 (2010)
10. Li, Y., Zhang, Z., Wang, C., Zhao, W., Chen, H.-H.: Blind cooperative communications for multihop Ad Hoc wireless networks. IEEE Trans. Veh. Technol. **62**(7), 3110–3122 (2013)
11. Kotteswari, K., Manikandan, S.P., Manimegalai, R.: QuAADD: a quick access routing algorithm using distance and direction of nodes in MANET. Int. J. Appl. Eng. Res. **10**, 1011–1022 (2015). ISSN 0973-4562
12. Merkel, S., Mostaghim, S., Schmeck, H.: Hop count based distance estimation in mobile adhoc networks-challenges and conseguences. Ad Hoc Netw. **15**, 39–52 (2014)
13. Son, T.T., Minh, H.L., Sexton, G., Aslam, N.: A novel encounter-based metric for mobile ad-hoc networks routing. Ad Hoc Netw. **14**, 2–14 (2014)
14. Wang, Z., Zekavat, S.A.: A novel semidistributed localization via multinode TOA–DOA fusion. IEEE Trans. Veh. Technol. **58**(7), 3426–3435 (2009)
15. Ciurana, M., Barceló-Arroyo, F., Martín-Escalona, I.: Comparative performance evaluation of IEEE 802.11v for positioning with time of arrival. Comput. Stand. Interfaces **33**(3), 344–349 (2011)
16. Senthil Kumaran, T., Sankaranarayanan, V.: Early congestion detection and adaptive routing in MANET. Egypt. Inf. J. **12**(3), 165–175 (2011)
17. Guo, L., Zhang, L., Peng, Y., Wu, J., Zhang, X., Hou, W., Zhao, J.: Multi-path routing in Spatial Wireless Ad Hoc networks. Comput. Electric. Eng. **38**(3), 473–491 (2012)
18. Subhan, F., Hasbullah, H., Ashraf, K.: Kalman filter-based hybrid indoor position estimation technique in bluetooth networks. Int. J. Navig. Obs. **2013**, 1–13 (2013)

A Novel Coherence Particle Swarm Optimization Algorithm with Specified Scrutiny of FCM (CPSO-SSFCM) in Detecting Leukemia for Microscopic Images

A. R. Jasmine Begum[1](✉) and T. Abdul Razak[2](✉)

[1] Cauvery College for Women, Trichy, Tamil Nadu, India
cauverycollege_try@rediffmail.com
[2] Jamal Mohammed College, Trichy, Tamil Nadu, India
abdull964@yahoo.com

Abstract. Image segmentation is a much opted technique in the image processing arena. Fuzzy C-Means clustering method has been widely used for medical image segmentation. In the Standard FCM the cluster centers are chosen randomly, which may lead to the dismal performance of clustering. In order to overcome the drawback of the FCM, this paper proposes a Novel Coherence Particle Swarm Optimization Algorithm with Specified Scrutiny of Fuzzy C-Means (CPSO-SSFCM). In this work, the Standard Fuzzy C-Means algorithm is fine-tuned using Particle Swarm Optimization algorithm to find the optimal cluster heads for segmentation of the White Blood Cells. Following the segmentation of nucleus and cytoplasm regions, the proposed algorithm is applied for the classification and optimisation of the result using Support Vector Machine. The values obtained from this method are compared with the Standard FCM, HFCMCCE and EHFCMCCE using quality parameters like Full Reference pixel based measures PSNR, MSE and statistical measures such as sensitivity, specificity and accuracy.

Keywords: FCM · PSNR · MSE · Support Vector Machine · Sensitivity
Specificity · Accuracy

1 Introduction

The Segmentation is the process that subdivides an image into its constituent parts or objects. Accurate segmentation of objects of interest in an image further improvises the analysis of these objects. Much desired segmentation techniques are edge detection and clustering techniques. Unsupervised clustering is required to identify the interesting patterns or groupings in a given set of data. In the area of pattern recognition of an image processing, the unsupervised clustering is used for "segmenting" the images and it can be a very effective technique to pinpoint natural groupings in data from a large data set, thereby letting concise representation of relationships embedded in the data [1].

Leukemia has become a modern day curse on mankind and turned to be a deadly disease claiming lives of thousands. Leukemia is a medical condition, where the

© Springer Nature Singapore Pte Ltd. 2018
G. Ganapathi et al. (Eds.): ICC3 2017, CCIS 844, pp. 58–73, 2018.
https://doi.org/10.1007/978-981-13-0716-4_6

combination of cancers initiated from the bone marrow and ends in forming excessive numbers of abnormal white blood cells. These abnormal blood components are termed as leukemia cells or "blasts". Based on the rate of leukemia cells growth condition, it is termed as acute or chronic. The type of cells infected by Leukemia defines its condition – myelocytes or lymphocytes and, leukemia is classified as lymphocytic, chronic myelocytic and acute myelocytic.

There is an urgent requirement of detecting leukemia automatically as the present system involves manual methods examining the blood smear as the first step toward diagnosis, the method is a lengthy process and it's accuracy is unproven. Morphological, textural and color features are extracted from the segmented nucleus and cytoplasm regions of the leukemia images, which facilitates hematologists towards timely identification and detection of leukemia from blood microscopic images, which could save the patient life. From the close review of allied work and published materials, it could be seen that high number of researchers utilized wide range of segmentation strategies [2].

Saha et al. [3] has proposed a novel and robust method for segmenting nucleus in overlapping Pap smear images, which forms a basis for further processing of cell images. The method proven to be successful in detecting and segmenting nucleus for isolated, touching and overlapping cells in Pap smear images and it also evolves a circular shape function with FCM clustering to improvise the image data partition. Dice coefficient, precision and recall were used for quantitative evaluation of the proposed method.

Gu et al. proposed [4] a new clustering algorithm, called sparse learning based fuzzy c-means (SL_FCM). Firstly, to reduce the computation complexity of the SR based FCM method and removing the redundant information in the discriminant feature is also done to improve the clustering quality. This algorithm performs better than other state-of-art methods with higher accuracy, for the large scale dataset and image.

Ananthi et al. [5] developed a new threshold based segmentation technique cantered on Interval-Valued Intuitionistic Fuzzy Sets (IVIFS). This methodology was developed to sort out the issue of selecting the values of membership function to symbolize imprecise data. The main aim of this paper is to segment leukocytes in blood smear images with the help of IVIFSs. An IVIFS is to be identified among the 256 IVIFSs having maximizing ultra-fuzziness along with varying threshold.

Liu et al. [6] introduce a first time fluid identification method in carbonate reservoir based on the modified Fuzzy C-Means (FCM) Clustering algorithm. Both initialization and globally optimum cluster center are generated from Chaotic Quantum Particle Swarm Optimization (CQPSO) algorithm, which is prudently used to evade the drawback of sensitivity to initial values and simply falling into local convergence in the Standard FCM Clustering algorithm.

A dynamic niching clustering algorithm based on individual-connectedness (DNIC) has been developed by Chang et al. [7] in which he proposes a compact k-distance neighborhood algorithm and an individual-connectedness algorithm. The algorithm derives the adaptive selection of the number of the niches to and dynamically identifies the niches. Many of the datasets with varying cluster volumes along with noisy points is successfully processed using DNIC clustering algorithm.

Shang et al. [8] proposed a clone kernel spatial FCM (CKS_FCM), which enhances segmentation performance through the generation of initial cluster centers, and by combining spatial information into the objective function of FCM and utilized a non-Euclidean distance based on a kernels metric, in place of the Euclidean distance traditionally used in FCM.

Jasmine Begum et al. [9] proposed a Hybrid Fuzzy C-Means Algorithm with cluster center Estimation (HFCMCCE), which hybridises the FCM with subtractive clustering for Leukemia Image Segmentation. This method is found to be suiting even for the images with 90% noise density up to 90%, additionally there is an increase in PSNR value and reduction in the Mean Squared Error for HFCMCCE applied image corresponding to the input and FCM applied image.

E. Rajaby, developed a novel method for color image segmentation by using only hue and intensity components of image and combines those by adaptive tuned weights in a specially defined fuzzy c-means cost function. This method specifies proper initial values for cluster centers with the aim of reducing the overall number of iterations and avoiding converging of FCM to wrong centroids. This algorithm showing a better performance is segmentation and speed compared to the other similar methods.

A Study is proposed by Jose L. Salmeron using a well-known soft computing method called Fuzzy Cognitive Maps (FCMs) for the early diagnosis of Rheumatoid Arthritis (RA) in order to assist physicians. Then, Particle Swarm Optimization (PSO) and FCMs along with medical experts' knowledge were used to model this problem and calculate the severity of this disease. The obtained result shows that this tool will be useful for General Practitioner's (GPs) to timely diagnosis of patients with RA.

The prime objective of this paper is to develop an advanced methodology in medical diagnosis bettering the available regular and other ambiguous approaches. Supplemental to the above objective, the paper proposes a computer-aided diagnosis system to assist the doctors in assessing medical images for diagnosing the disease in leukemia patients at the earlier stage.

2 Standard Fuzzy C-Means Algorithm

Fuzzy C-Means was initially proposed by Bezdek et al. It is the widely used tool for image processing in clustering objects in an image. FCM facilitates the pixels to secure a place with various cluster along with alterable degrees of participation. Owing to this extra adaptability, FCM is also termed as Soft clustering strategy. But in hard clustering, the data gets portioned into a specified number of mutually exclusive subsets. Fuzzy clustering is a simple methodology when compared to the hard clustering and it carries out the non-unique partitioning of the data in a collection of clusters [10].

The Standard Fuzzy C-Means Algorithm is as follows:

Step1: Randomly initializing the cluster centers, termination criteria α, Maximum no of iterations X.

Step 2: Creating a distance matrix from a point x_j to each of the cluster centers using the following equation.

$$d_{ij} = \left\| c_i - x_j \right\| \tag{1}$$

Step 3: Repeat the following steps until reach the total number of iterations.

Step 4: Compute the Membership matrix.

$$u_{ij} = \frac{1}{\sum_{k=1}^{c} \left[\frac{d_{ij}}{d_{kj}}\right]^{2/(m-1)}} \tag{2}$$

Step 5: Generating new cluster centers.

$$c_i = \frac{\sum_{j=1}^{n} u_{ij}{}^m x_j}{\sum_{j=1}^{n} u_{ij}{}^m} \tag{3}$$

Step 6: Compute the objective function.

$$J = \sum_{j=1}^{N} \sum_{i=1}^{c} (\mu_{ij})^m d_{ij}{}^2 \tag{4}$$

Step 7: Update cluster heads.

Step 8: If abs value of distance metric of J is $< \alpha$ Stop execution.

Step 9: Otherwise update objective function values and Go to Step 3.

---

3   Particle Swarm Optimization

The particle swarm optimization (PSO) is proposed by Eberhart and Kennedy. PSO is a simple yet a robust search technique; it is used in a range of search and optimization problems, which includes image processing problems such as image segmentation. The simple explanation of the PSO is that the swarm of birds looks for its feed reiteratively the region around the bird, which seems to be near the food mostly and gets its feed at the end. After the repetitive process, particles traverse towards the region around the and G_{Best}, so that it could arrive at the optimal point. Since each particle consist of several component particles, the process of particles movements is essentially that the component particles traverse in the direction of the corresponding particles of the pBest and gBest particles. So it becomes obvious that if there exists no corresponding relationship between the component particles of each particle the evolution process of the particles will be disordered and could not converge to the optimal solution [11, 12].

$$V_i(t+1) = \omega.V_i(t) + c_1.r_1.(X_i^l(t) - X_i(t)) + c_2.r_2.(X^g - X_i(t)) \tag{5}$$

$$X_i(t+1) = X_i(t) + V_i(t+1) \tag{6}$$

ω is the inertial weight.

$V_i(t)$ is the previous velocity in iteration t of ith particle.

c_1 and c_2 are coefficients.

r_1 and r_2 are random numbers ranging between 0 to 1.

$(X_i^l(t) - X_i(t))$ is the difference between the local best X_i^l of the ith particle and previous position $X_i(t)$.

$(X^g - X_i(t))$ is the difference between the global best X^g and previous position $X_i(t)$.

The Particle Swarm Optimization Algorithm as follows

Step 1: Initialize swarm, which includes the population size, starting position and velocity of particles, etc.

Step 2: Calculate fitness for each particle storage each particle best position P_{Best} and its fitness value. Choose the particle that has the best fitness as G_{Best}.

Step 3: Update each particle's velocity V and position X according to the equation 5 and equation 6.

Step 4: Calculate the fitness of each particle after update the position. Compare the fitness of each particle with its best previous fitness P_{Best} if better than it then set the current position as P_{Best}.

Step 5: Compare the fitness of each particle with the group best previous fitness if better than it then set the current position as G_{Best}.

Step 4: Check if the algorithm meets the condition like maximum number of iterations or acceptable fitness of the G_{Best} or tolerable convergence of all particles then stop and output the optimal solution. If conditions not met then go to step3.

3.1 Coherence Particle Swarm Optimization with Specified Scrutiny of Fuzzy C-Means (CPSO-SSFCM) Algorithm

The various stages involved in the proposed algorithm CPSO-SSFCM include enhancement of microscopic images, segmentation of background cells, features extraction, and finally the classification. The Figs. 1 and 2 shows the Flow diagram and schematic diagram of the proposed algorithm. The Proposed Algorithm CPSO-SSFCM is given as follows:

Segmentation Phase-CPSO-SSFCM:

Input: Normal and Leukemia infected blood cell images

1. Initialize cluster numbers, Let C=4.
2. Initialize random cluster centre value.
3. Initialize α, initialize maximum no of iterations X.
   X=10, α =0.000001
4. Inner product norm metric is chosen (such as distance metric)
5. Initialize membership matrix using equation 2.
6. Calculate new cluster centre using equation 3.
7. Do update membership matrix and the objective function (J) using equations 2 and 4 respectively.
8. If abs value of distance metric of J is < α Stop execution
9. Otherwise update objective function values Go to Step 10.
10. Initialize the population size.
11. Compute velocity and position using equation 5 and 6 respectively to calculate P_{Best} (**Personal best**).
12. If position > P_{Best} then Best position is P_{Best}.
13. If P_{Best} < G_{Best}(Global Best) then P_{Best} is the G_{Best}.
14. Update cluster head with the G_{Best}.
15. If max number of iterations not arrived go to Step 6.

Output: Segmented White Blood Cell (WBC) Image

Feature Extraction Phase CPSO-SSFCM:

Input: The Segmented WBC Image

1. Let N is the no of input images.

2. Using Canny Edge Detection extract the Nucleus (N_{ext}). and Cytoplasm ($C_{ext(i)}$) of the clustered Image Cl_i.

3. Compute Region properties of Rp ($N_{ext}(i)$) and Rp ($C_{ext(i)}$).

4. Stop the execution.

Output: Extracted Nucleus and Cytoplasm of segmented WBC Image

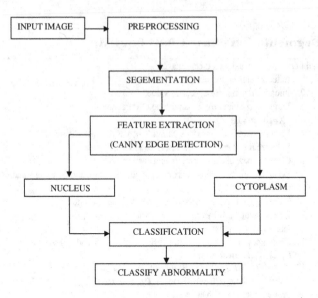

Fig. 1. Schematic diagram of the proposed algorithm

4   Result Analysis

In order to validate the effectiveness of the method proposed in this paper, using coherence particle swarm optimization with Fuzzy C-Means Clustering algorithm based on the direct histogram to detect the normal and leukemia infected blood microscopic images.

The proposed method is implemented using the Image Processing Tools of MATLAB R2013a. The blood smear images were collected from dataset source [16]. These images were digitalized with the digital camera connected to a Carl Zeiss photo microscope with a magnification of 200x. The size of image is 256 × 256. The test is conducted on 50 images and in this paper detection of Acute Myelocytic Leukemia (AML) affected image and normal image is shown. The results of the proposed algorithm CPSO-SSFCM is compared with the standard FCM, HFCMCCE [8], EHFCMCCE [17] based on Full Reference and pixel based Image quality measures such as PSNR, MSE and statical measures such as Sensitivity, Specificity and Accuracy.

4.1   Pre-processing Step

For image denoising, linear filters are less effective to remove the noise. The best solution is to use nonlinear filters like median filter. The median filtering process is executed by replacing the central pixel with the median of all the pixels value in the current neighbourhood [18]. The Fig. 3 illustrates the pre-processing stage where the median filter is applied on the input image to remove the noise to equalize the gravy levels of image intensities.

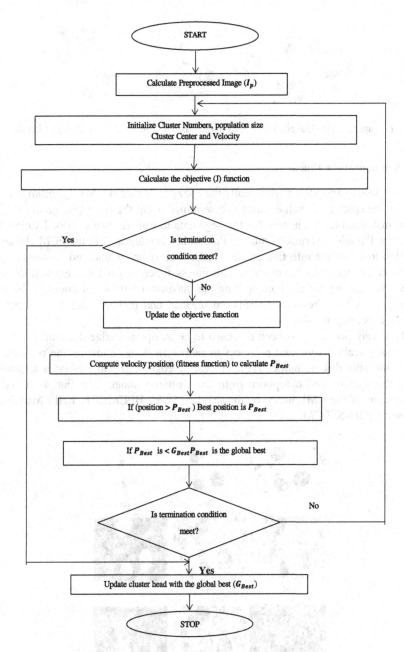

Fig. 2. The flow chart for the proposed algorithm

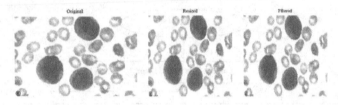

Fig. 3. (a) Input image (leukemia infected) (b) Resized to 256 × 256 (c) Median filtered image

4.2 Segmentation Phase

In the proposed CPSO-SSFCM, initially the Fuzzy C-Means (FCM) algorithm is tuned based on the spatial hue value which is determined using the color histogram to detect the optimal number of clusters for the segmentation of the White Blood Cells. The Coherence Particle Swarm Optimization algorithm is integrated to the FCM clustering algorithm to select the effective cluster heads. Cluster heads undergo crossover operation with the nodes in the clusters. The fitness function get the minimum distance between the cluster nodes by comparing the maximum number of nodes in the each cluster, from which the cluster heads are selected and updated each time to get the effective clustering results.

The Canny operator has been designed to be an optimal edge detector. It takes as input a grey scale image, and produces as output an image showing the positions of tracked intensity discontinuities [19]. The canny edge detection method is applied to extract the nucleus and cytoplasm from the clustered image. The Fig. 4 shows the segmentation of the AML image by the standard FCM, HFCMCCE, EHFCMCCE and proposed CPSO-SSFCM.

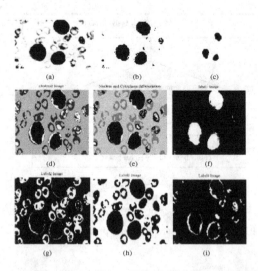

Fig. 4. (a) Standard FCM image (b) HFCMCCE image (c) EHFCMCCE image (d) CPSO-SSFCM image (e) Nucleus and cytoplasm differentiation (f) Label1 image (g) Label2 image (h) Label3 image (i) Label4 image

4.3   Feature Extraction Phase

The most important task in pattern recognition is selecting the proper diagnostic features, describing the image by the numerical values, and enabling the automatic system to perform the recognition. In this step the geometrical and size ratio features from the regions like nucleus cytoplasm and the WBC is extracted.

Geometric Features

(i)   **Area:** It is evaluated by calculating all nonzero pixels within image region.

$$Area = \sum_{i=1}^{n} \sum_{j=1}^{m} b_{ij} \tag{7}$$

where b_{ij} is the value of binary image (0,1) at the pixel coordinate (i, j) with in a m × n image.

(ii)   **Perimeter:** The perimeter was measured by computation distance between the Successive boundary pixels.

(iii)   **Circularity:** Area-to-perimeter ratio is the measure of roundness or circularity But local irregularities are not reflected by this feature. It is defined as:

$$Circularity = 4 * \pi * Area \,/\, perimeter^2 \tag{8}$$

A circle gets the value of 1, while objects with bumpy boundaries get lower values.

(iv)   **Eccentricity:** This parameter is used to measure to what extend the shape of a nucleus deviates from being circular. It is an important feature since Monocytes are more circular than the Monoblast. The value of eccentricity ranges between 0 and 1. If the value of the eccentricity is below one then it is not circular. Eccentricity is provided by the Eq. (9).

$$Eccentricity = \sqrt{a^2 - b^2} \,/\, a \tag{9}$$

Where a is the semi minor axis and b is the semi major axis.

(v)   **Solidity:** The solidity is the ratio of actual area and the convex hull area and is also an essential feature for classification of a blast cell. This measure is defined in Eq. (10) [20].

Size Ratio Measure

(i)   **Nucleus to Cytoplasm Ratio (NCR):** It is a ratio of the area of the nucleus to the area of the cytoplasm. It is a measurement to indicate the maturity of a cell, because as a cell matures, the size of its nucleus generally decreases. Pre-cancerous cells have increased nucleus to cytoplasm ratio. Malignant cells occur in clumps and have irregularly shaped nuclei and cytoplasm [21].

$$NC\ Ratio = \text{Nucleus area} \ / \ \text{Cytoplasm area} \qquad (10)$$

The geometric feature of the WBC is one of the prime factor to validate whether the segmented WBC can be treated as either normal or leukaemia infected. The experimentation is conducted on 40 leukemia infected and 10 normal blood microscopic images. The area, radius and diameter of the leukemia affected input image is shown in the Table 1. The Eqs. (11)–(13) are used to calculate the diameter and radius of the WBC in pixels which is then converted into micrometer (μm).

The Fig. 5 shows the subimage of the WBC. The diameter of the normal monocytes is 12–20 μm [21]. But, for the input image the diameter is greater than the normal size.

$$WBC\ Radius = \sqrt{\frac{Area}{\pi}} \qquad (11)$$

$$WBC\ Radius = WBC\ Diameter \ / \ 2 \qquad (12)$$

$$WBC\ Diameter = 2 * \sqrt{\frac{Area}{\pi}} \qquad (13)$$

According to the hematologist the shape of the nucleus and cytoplasm is an essential feature for distinguish of blasts. The geometric features are extracted for shape analysis of the nucleus and cytoplasm. The extracted sub image of nucleus, cytoplasm of the leukemia infected and normal blood cell image using bounding box technique is given in the Fig. 6.

Fig. 5. WBC sub image (a) Cell 1 (b) Cell 2 (c) Cell 3

Table 1. Geometric features of WBC

Measure	Cell 1	Cell 2	Cell 3
WBC Area (in pixels)	3694	2198	3377
WBC Radius (in μm)	12.32	10.33	11.78
WBC Diameter (in μm)	24.64	20.66	23.56

The extracted shape features Area, Perimeter, Eccentricity, Circularity, Solidity and Nucleus to Cytoplasm Ratio of Nucleus 1, 2 and 3 is shown in Table 2. The circularity value is less than 0.90, which indicates that the nucleus is not circular and also denotes the distortion in the shape of the nucleus. If eccentricity value is near to one indicates that the nucleus is a blast. Considering the solidity, if its value is below one then it seems to possess irregular boundaries which again substantiates that the input image contains a blast of the nucleus. As Pre-cancerous cells have increased nucleus to cytoplasm ratio here the NCR value of the Nucleus 1, 2 and 3 is high which in turn indicates it must be a blast.

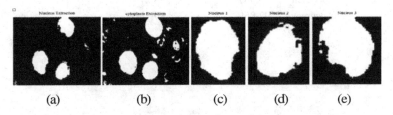

      (a)   (b)   (c)   (d)   (e)

Fig. 6.   (a) Nucleus extracted image (b) Cytoplasm extracted image (c) Sub image Nucleus 1 (d) Sub image Nucleus 2 (e) Sub image Nucleus 3

Table 2. Geometric and size ratio features of nucleus and cytoplasm [Blast - Eccentricity $\approx$ 1, Solidity < 1 and Circularity < 0.90 & NCR > 1]

Measure	Area	Perimeter	Eccentricity	Circularity	Solidity	Nucleus to Cytoplasm Ratio (NCR)
Nucleus 1	3154	262.5513	0.8688	0.5757	0.9189	5.8407
Nucleus 2	1352	175.1543	0.4270	0.5535	0.8977	1.5981
Nucleus 3	2853	231.9239	0.7642	0.6661	0.9510	5.4447

4.4   Classification Phase

Support Vector Machine (SVM), proposed by Vapnik, is a well-known pattern recognition tools and was associated with various areas such as text mining, bioinformatics, image classification, cancer diagnosis, and feature selection [18]. After feature extraction of the nucleus and cytoplasm the SVM classification is performed to classify the input image as normal or abnormal.

To ensure the effectiveness of the classifier and segmentation the following parameters are calculated.

True Positive (TP) = No of images having leukemia and detected

True Negative (TN) = No of images that have not leukemia and not detected

False Positive (FP) = No of images that have not leukemia and detected

False Negative (FN) = No of images have leukemia and not detected [18].

The Fig. 7 shows the classification of the input image as AML. The Table 3 shows the results of the proposed algorithm for SVM classifier with values 100.00%, 92.85% and 83.33%, for sensitivity, accuracy, and specificity, respectively which is exhibiting a good performance of the functioning of the proposed method.

Table 3. Performance analysis of proposed method using SVM classifier

Parameters	CPSO-SSFCM image	EHFCMCCE image
Sensitivity	100.00%	81.25%
Specificity	83.33%	83.27%
Accuracy	92.85%	89.79%

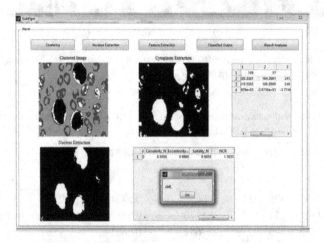

Fig. 7. Classification of input image as AML

Table 4. Performance analysis of full reference and pixel difference based measure on existing and proposed algorithm [Ideal values: PSNR-High, MSE-Low]

PERFORMANCE ANALYSIS OF STANDARD FCM, HFCMCCE,EHFCMCCE and CPSO-SSFCM BASED on PSNR AND MSE		
File Edit View Insert Tools Desktop Window Help		
	PSNR(dB)	MSE
STANDARD FCM	26.4200	29.78
HFCMCCE	34.0800	20.73
EHFCMCCE	34.4400	18.17
CPSO-SSFCM	36.2900	15.25

Finally the performance of the proposed method is analysed by calculating the Full Reference and Pixel difference based Measures such as Peak Signal to Noise Ratio (PSNR) and Mean Square Error (MSE) where PSNR of the image is measured by the ratio between the maximum possible power of an image and a power of corrupting noise and higher the PSNR better the quality of the image [17]. MSE is used to measures gray-level difference between pixels of the ideal and the distorted images [22]. From the Table 4 and the comparison chart given in the Fig. 8, it can recognized that the proposed method is showing high PSNR and low MSE values which reveals that the segmentation using the proposed method is showing good results than the Standard FCM, HFCMCCE [8] and EHFCMCCE [18].

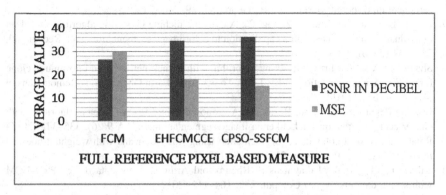

Fig. 8. Performance analysis of full reference and pixel difference based measure on existing and proposed algorithm

5 Conclusion

The obtained results confirms that the proposed algorithm CPSO-SSFCM, which is developed for leukemia image segmentation performs well in comparison to the regular and existing methods used in practice. The cluster heads are optimized by utilizing CPSO. The cluster heads are converged within 4 number of iterations which asserts the efficiency of the proposed algorithm. Then the segmented WBC, nucleus and cytoplasm is analysed based on the Geometric and size ratio measures. Immediately, the classification is made to detect and classify the normal and leukemia infected images. The results of SVM classifier is examined based on the Sensitivity, Specificity, Accuracy and ROC curve. Finally the effectiveness of the algorithm is assessed on the Full Reference and pixel based Reference measure such as PSNR and MSE. As a future work, the algorithm can be extended to identify the sub type of the leukemia and in the analysis phase additional measures such as statistical, textural can be presented.

References

1. Gonzalez, R.C., Woods, R.E.: Digital Image Processing, 2nd edn. Prentice-Hall of India Pvt. Ltd., New Jersey (2002)
2. https://en.wikipedia.org/wiki/Leukemia. http://onlineessays.com/essays/tech/leukemia.php
3. Saha, R.: Spatial shape constrained fuzzy C-Means (FCM) clustering for nucleus segmentation in pap smear images, 978-1-5090-2896-2/16/$31.00 ©2016. IEEE
4. Gu, J.: Sparse learning based fuzzy c-means clustering. Knowl.-Based Syst. **119**, 113–125 (2017)
5. Ananthi, V.P.: A new thresholding technique based on fuzzy set as an application to leukocyte nucleus segmentation. Comput. Methods Programs Biomed. **134**, 165–177 (2016)
6. Liu, L.: A modified fuzzy C-Means (FCM) clustering algorithm and its application on carbonate fluid identification. J. Appl. Geophys. **129**, 28–35 (2016)
7. Chang, D., Zhao, Y., Liu, L., Zheng, C.: A dynamic niching clustering algorithm based on individual - connectedness and its application to color image segmentation. Patt. Recogn. **60**, 334–347 (2016)
8. Shang, R.: A spatial fuzzy clustering algorithm with kernel metric based on immune clone for SAR image segmentation. IEEE J. Sel. Topics Appl. Earth Observations Remote Sens. **9**(4), 1640–1652 (2016)
9. Jasmine Begum, A.R., Abdul Razak, T.: A proposed hybrid fuzzy C-means algorithm with cluster center estimation for Leukemia image segmentation. IJCTA **9**(26), 335–342 (2016)
10. Rajaby, E.: Robust color image segmentation using fuzzy c-means with weighted hue and intensity. Dig. Sig. Procession **51**, 170–183 (2016)
11. Salmeron, J.L.: Medical diagnosis of Rheumatoid Arthritis using data driven PSO–FCM with scarce datasets. Neuro Comput. **232**, 104–112 (2017)
12. Chapter4, Fuzzy Clustering. https://homes.di.unimi.it/valenti/SlideCorsi/…/Fuzzy-Clustering-lecture-Babuska.pdf
13. Particle Swarm Optimization applied to Image Vector Quanitzation. http://books.google.co.in/books
14. http://www.dii.unipd.it/ ~ alotto/didattica/corsi/Elettrotecnica%20computazionale/pso.pdf
15. https://www.pantechsolutions.net/image-processing-projects/matlab-code-for-image-retrieval
16. http://www.me.chalmers.se/ ~ mwahde
17. Ding, Z., Sun, J., Zhang, Y.: FCM image segmentation algorithm based on color space and spatial information. Int. J. Comput. Commun. Eng. **2**(1), 48–51 (2013)
18. Atlas of hematology. http://www.hematologyatlas.com/leukemias.htm
19. Jasmine Begum, A.R., Abdul Razak, T.: The performance comparison of spatial filtering based on the full reference image quality measures PSNR, RMSE, MSSIM and UIQI in medical image improvement. Int. J. Appl. Eng. Res. (IJAER) **10**(82), 97–102 (2015). ISSN 0973-562
20. Jasmine Begum, A.R., Abdul Razak, T.: A proposed novel method for detection and classification of Leukemia using blood microscopic images. Int. J. Adv. Res. Comput. Sci. (IJARCS), **8**(3) (2017)
21. Jasmine Begum, A.R., Abdul Razak, T.: Segmentation techniques: a comparison and evaluation on mr images for brain tumour detection. Int. J. Adv. Res. Comput. Sci. **7**(2) (2016)

22. Pawar, M.P.K.: A survey on analysis of malignant cervical cells based on N/C ratio. Int. Res. J. Eng. Technol. (IRJET), **03**(06) (2016)

23. http://www.cap.org/apps/docs/proficiency_testing/2012_hematology_glossary.pdf

24. Zhang, K., Wang, S., Zhang, X.: A new metric for quality assessment of digital images based on weighted-mean square error. Proc. SPIE **4875**, 1–6 (2002)

Impact of Negative Correlations in Characterizing Cognitive Load States Using EEG Based Functional Brain Networks

M. Thilaga[1(✉)], R. Vijayalakshmi[2], R. Nadarajan[1],
and D. Nandagopal[3]

[1] Computational Neuroscience Laboratory, Department of Applied Mathematics
and Computational Science, PSG College of Technology,
Coimbatore, Tamil Nadu, India
thilagaselvan@gmail.com
[2] Department of Computer Science and Software Engineering,
Miami University, Oxford, USA
rvpsgtech@gmail.com
[3] Cognitive Neuroengineering Laboratory, School of Information Technology
and Mathematical Sciences, Division of IT, Engineering and the Environments,
University of South Australia, Adelaide, Australia

Abstract. The human brain is one of the least understood large-scale complex systems in the universe that consists of billions of interlinked neurons forming massive complex connectome. Graph theoretical methods have been extensively used in the past decades to characterize the behavior of the brain during different activities quantitatively. Graph, a data structure, models the neurophysiological data as networks by considering the brain regions as nodes and the functional dependencies computed between them using linear/nonlinear measures as edge weights. These functional connectivity networks constructed by applying linear measures such as Pearson's correlation coefficient include both positive and negative correlation values between the brain regions. The edges with negative correlation values are generally not considered for analysis by many researchers owing to the difficulty in understanding their intricacies such as the origin and interpretation concerning brain functioning. The current study uses graph theoretical approaches to explore the impact of negative correlations in the functional brain networks constructed using EEG data collected during different cognitive load conditions. Various graph theoretical and inferential statistical analyses conducted using both negative and positive correlation networks revealed that in a functional brain network, the number of edges with negative correlations tends to decrease as the cognitive load increases.

Keywords: EEG · Functional Brain Networks · Cognition · Graph theory
Correlation

© Springer Nature Singapore Pte Ltd. 2018
G. Ganapathi et al. (Eds.): ICC3 2017, CCIS 844, pp. 74–86, 2018.
https://doi.org/10.1007/978-981-13-0716-4_7

1   Introduction

One of the most complex biological structures in the earth, the human brain, has approximately 10^{11} neurons each with 10^4 synaptic connections that result in a massive complex connectome with one quadrillion cellular connections in total. These neurons process information parallelly and enable the human brain to work millions of times faster than any supercomputer in existence today. Understanding the dynamic interactions of the human connectome is an active area of research that requires efficient computational methods derived from different disciplines [4]. Neuroscience is a multidisciplinary scientific research field which investigates the complex dynamic nervous system of the brain. The goal of most of the neuroscientific researchers is to model the brain's dynamic interaction patterns thereby understand and analyze the brain functioning during different activities. Connectivity-based models derived from graph theory have gained significant attention in the past decades because it enables to visualize and investigate how individual elements (brain regions) are interacting with each other [1]. It also provides efficient techniques and measures to study the dynamic interaction patterns of the brain that generates emergent behaviors during a specific activity and to identify the most central brain regions affected by these behaviors. Moreover, by modeling the brain's interactions as a network of interconnected regions, it is possible to study and understand the higher order processes such as cognition and the progression of the disease. Numerous studies in the literature of brain network analysis have addressed the problem of uncovering the patterns for characterizing impairments in cognitive activities. Nonetheless, analyzing the healthy brain to understand the underlying neuronal functions is also essential to identify cognitive impairments causing various mental health issues [5, 6].

Biomedical science has been a growing field of research intended to store, analyze, and visualize massive time series datasets. This maturity is due to the advancements, most notably in the past five decades, in digitizing and storing voluminous data recorded using various neuroimaging modalities (such as *MRI*, fMRI, and *PET*) and neurophysiological recordings (such as *EEG* and *MEG*). Devising efficient computational techniques to understand how the interactions among billions of neuronal elements result in cognition is one of the most challenging problems in the field of computational neuroscience. Although various advanced techniques are available to capture the electrical activity of the brain, EEG, despite being the oldest technology, is still used in many clinical studies and research due its high temporal resolution, noninvasive nature, low hardware cost, and ease of use [3].

Some studies on brain network analysis in the literature have used Pearson's correlation coefficient, one the most basic assessment measures of functional connectivity, to compute the linear dependence between two brain regions [16, 18–20]. The correlation value is then used to characterize the connection strength between the brain regions in Functional Brain Networks (*FBNs*). Positive correlation values between the brain regions indicate that the activity in one region increases the activity in the other region. On the contrary, the negative correlation indicates that the activity in one region decreases the activity in another region (antagonistic) [17]. Higher the correlation value, stronger is the relationship between the brain regions [10, 11]. Many of the existing studies on *FBN* analysis have considered only the edges with positive correlation values as their edge weights ignoring the negative ones from the analysis. Negative correlation remains a topic of debate concerning the neurophysiological function of the brain and interpretation [12]. Also, there is insufficient literature investigating the relationships between the brain regions with negative correlations. This study investigates the impact of the negative correlations in characterizing different cognitive load states of the brain.

The remainder of the paper is structured as follows. Section 2 provides an overview of FBN analysis using graph-theoretic approaches and negative functional connectivity in network analysis literature. Section 3 describes the experimental setup used for data acquisition and preprocessing methods. The proposed framework to analyze the impact of negative correlations in characterizing different cognitive load conditions is presented in Sect. 4. A detailed discussion of the experimental results is presented in Sect. 5. Section 6 provides summary of findings including future directions for this research.

2 Functional Brain Network Analysis Using Graph Theory

There has been a growing popularity of modeling brain connectivity using graphs in the past decades to understand the dynamic patterns of electrical activity of neurons during various activities. Graph data structure has been widely used in many applications to assess the topological properties since it represents the relationships between the entities efficiently. Application of graph theoretical analysis to neuroimaging data enables to develop a quantitative framework to describe and understand the brain functioning. Typically, a graph G is defined as a collection of vertices (V) and edges (E), $G = (V, E)$. Brain connectivity is modeled as a graph by considering the brain regions and the physical connections/functional associations among all the brain regions as nodes and edges respectively [1, 2, 7–9]. The *FBNs* are acquired from neurophysiological data by estimating the synchronization between the electrode pairs using linear or nonlinear statistical measures such as Pearson's correlation coefficient, mutual information, etc. The distributed nature of cognitive activity has been understood through conceptualizing different cognitive processes as *FBNs*. In this study, r, which is a measure of the linear relationship between two variables in terms of the covariance of the variables normalized to their standard deviations is used for constructing *FBNs* [13]. Given two signals X and Y, the r value is computed as

$$r = \frac{\sum_{i=1}^{n}\left((x_i - \bar{x})(y_i - \bar{y})\right)}{\sqrt{\sum_{i=1}^{n}(x_i - \bar{x})^2 \sum_{i=1}^{n}(y_i - \bar{y})^2}} \tag{1}$$

where n is the sampling rate (per second) of the signals. The r value, calculated using X and Y, represents the degree of correlation of the signals in the range -1 to $+1$ inclusive. This study focuses more on studying the impact of negative correlation networks during mild and strong cognitive load states.

2.1 Negative Links in Network Analysis

In some of the application domains such as online social networks and brain networks, the interactions among the entities result in the formation of positive or negative relations. In social network data, the relationship between any two actors is represented by a link which may be either positive or negative. While the positive links between the entities represent friendship, endorsement, etc., the negative links represent opposition, distrust, and avoidance. For instance, in Slashdot, users are allowed to tag other users as friends and foes. Majority of the existing research works on online social network analysis have considered only the positive relationship between the actors. To consider both the positive and the negative relationships that exist between the entities in online social networks for analysis, they are modeled as signed networks in which the nodes represent the users, and the edges between them represent the positive or negative relationships depending on the nature of the interaction. An undirected signed graph G with V vertices and E edges is defined as a graph $G = (V, E, \sigma)$, where $\sigma: E \rightarrow \{-1, +1\}$ is the sign function [21]. Signed graph analysis methods have been applied to study and analyze different types of interaction patterns.

In brain network analysis, the *FBNs* constructed using linear correlation measures contain both positive and negative connections. The negative connections between brain regions are not generally considered for experimental analysis by a significant fraction of studies on *FBNs* due to many reasons such as (i) the neuroscientific interpretation of the negative correlations between brain areas is still unclear, (ii) to avoid uncertainty, (iii) to reduce computational costs, and (iv) the negative correlations are considered as less reliable connections than positive ones [12, 14]. While some authors state that negative correlations have neurobiologically meaningful interpretations, some researchers have considered ignoring the negatively correlated links from *FBNs* as a thresholding method that retains only the strongest links in the resulting thresholded subnetworks [15]. In this study, the positive and negative connections in the *FBNs* constructed using r are split into positive and negative correlation networks which are then used for further analysis. This paper is a pioneering attempt in using negative correlations of *FBNs* to characterize different cognitive load states of the brain.

3   Data Collection and Preprocessing

The EEG data used in this study is collected in the Cognitive Neuroengineering & Computational Neuroscience Lab (*CNEL*) at the University of South Australia, HREC Approval (30855), from nine participants P1 through P9. The details of the experimental setup, data collection, and data preprocessing steps can be referred from these publications [22, 23]. The data used for the experiments was acquired following the same procedure as the previous experiments by the authors. The preprocessed EEG data is used for the analysis of negative correlation in the brain network.

4 Characterizing Cognitive Load States of the Brain Using Negative Functional Connectivity Networks

The Negative Functional Connectivity Analysis framework shown in Fig. 1 is proposed to accomplish the objective of characterizing the interaction patterns of different cognitive load states of the brain based on negative correlation networks using a three-fold procedure.

(i)   Constructing a Graph Database (GD) from EEG data.
(ii) Constructing positive and negative functional connectivity networks.
(iii) Characterizing cognitive activity using positive and negative functional connectivity networks.

The following subsections are dedicated to describing the steps involved in modeling and analyzing the *FBNs* to characterize the cognitive behavior of the brain.

4.1 Graph Database Construction

The preprocessed EEG data (time domain-amplitude) of mild (*Drive*) and heavy cognitive load (*DriveAdo*) states are partitioned into a number of chunks (each of length two seconds). Each chunk of data is modeled as an *FBN* by considering the electrode sites (in this study 30 electrodes are considered) as nodes and the pair-wise functional associations of these sites measured using Pearson's correlation coefficient as weights on the edges. The resulting fully connected weighted undirected *FBNs* of mild and heavy cognitive load states are stored in *GD*.

4.2 Constructing Negative and Positive Correlation Networks

The *GD* consists of fully connected *FBNs* with either positive or negative correlation weights on the edges. The impact of negatively correlated edges in characterizing different cognitive load states of the brain is studied by splitting each FBN into two networks by including the positive weighted and negative weighted edges into Positive Functional connectivity (*PFC*) and Negative Functional Connectivity (*NFC*) networks respectively.

4.3 Analyzing *NFC* Networks

The *NFC* networks constructed using r are analyzed using complex network measures and statistical tests. First, the connectivity (edge) density using which any two networks

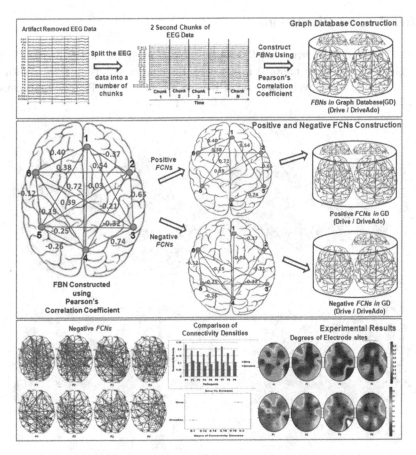

Fig. 1. Negative functional connectivity analysis framework

can be compared concerning connectedness is used for analysis. Connectivity density of a network is the ratio of actual number of connections that exists to the total number of possible connections. It is also called as physical cost or wiring cost, and its possible values lie between 0 and 1. A network with a connectivity density 1 means that the network is complete. The node degree (D) accounts for the number of connections a node has with its neighboring nodes. In a graph G with positive and negative weighted edges, the positive degree of a node u is the number of edges of node u having positive weights, i.e., $D^+(u) = |\{uv : uv^+ \in E\}|$, and the negative degree of node u, is $D^-(u) = |\{uv : uv^- \in E\}|$. Statistical techniques such as t-tests and one-way ANOVA are carried out to ensure that the mean differences between the connectivity densities and the degrees of nodes in *NFC* and *PFC* networks of *Drive* and *DriveAdo* states are significantly different. The following section is dedicated to presenting and discussing the results of the extensive experimental analysis.

5 Results and Discussions

The *NFC* networks of *Drive* and *DriveAdo* states showing the negative connections between the nodes (brain regions) in the networks constructed using *r* are shown in Fig. 2(a) and (b) respectively for four participants P1, P2, P3 and P4 (due to the space restriction).

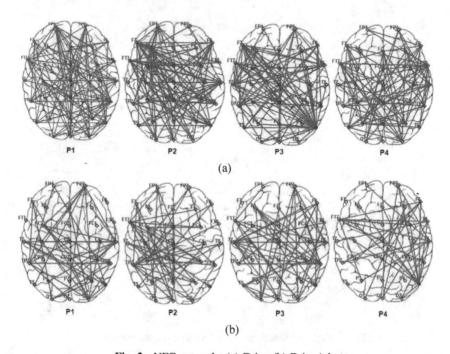

Fig. 2. NFC networks (a) Drive (b) DriveAdo

Firstly, the connectivity densities of these networks are computed to understand the topology of these networks at a macro level. The connectivity densities of both *NFC* and *PFC* networks of *Drive* and *DriveAdo* states of all the participants are shown in Fig. 3. It is interesting to note from the results of connectivity densities of the *NFC* networks of different cognitive load states that the *DriveAdo* network has relatively fewer negative connections when compared to *Drive* state. This indicates that there are many negatively correlated edges between the brain regions when the cognitive load is minimal. When the cognitive load increases, the negative correlation tends to decrease. In other words, many of the edges linking the electrode sites become positively correlated during *DriveAdo* state. The results of this analysis show that most of the brain regions exhibit highly cohesive interactions during *DriveAdo* state due to an increased cognitive activity resulting in an increased number of positive edges in the *FBN*.

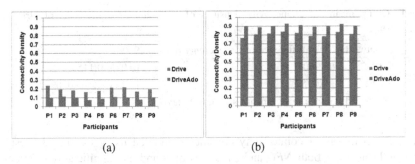

Fig. 3. Comparison of connectivity densities of *Drive* and *DriveAdo* states (a) *NFC* (b) *PFC* networks

The group means of connectivity densities of *NFC* and *PFC* networks of *Drive* and *DriveAdo* states are computed using t-test (two-tailed) at $\alpha = 0.05$, and the results are shown in Fig. 4(a) and (b) respectively. It can be observed from Fig. 4 that the means of connectivity densities of *Drive* and *DriveAdo* states computed using *NFC* and *PFC* networks are significantly different.

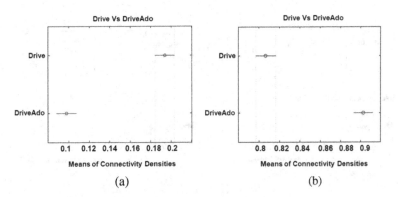

Fig. 4. Group means of connectivity densities of *Drive* and *DriveAdo* states (a) *NFC* (b) *PFC* networks

Table 1. Group means of connectivity densities of *NFC* and *PFC* networks of *Drive* and *DriveAdo* states

Network	Mean difference of connectivity density	95% CI
NFC	0.0960*	[0.076, 0.1151]
PFC	−0.0957*	[−0.1151, −0.0769]

*Mean difference is significant at $\alpha < 0.05$ level

The group means of connectivity densities computed using two-tailed t-test are shown in Table 1 for both *NFC* and *PFC* networks and are significantly different at 95% confidence interval. Therefore, the null hypothesis that the mean difference in the connectivity densities of mild and heavy cognitive load states is not significantly different is rejected at 5% significance level.

To evaluate the statistical significance of mean degree differences of the nodes in the *NFC* networks of *Drive* (*NFC_Drive*) and *DriveAdo* (*NFC_DriveAdo*) states and the *PFC* networks of these states (*PFC_Drive* and *PFC_DriveAdo*), one-way ANOVA test was performed with a 95% confidence interval. The multi-comparison tests performed on *NFC* and *PFC* networks of *Drive* and *DriveAdo* states of four participants are shown in Fig. 5. It is observed that the difference between the means of degrees of *NFC_Drive* and *PFC_Drive* is comparatively less than that the difference between the

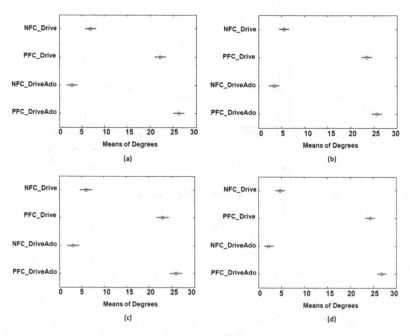

Fig. 5. Multi-comparison of means of degrees of nodes in *NFC* and *PFC* networks of Drive and DriveAdo states (a) P1 (b) P2 (c) P3 (d) P4

means of degrees of the *NFC_DriveAdo* and *PFC_DriveAdo* for all the participants. The difference in the mean degrees of *NFC* and *PFC* networks in *DriveAdo* state is significantly higher than the difference in *Drive* state. It becomes apparent from the analysis that the number of negative edges in an *FBN* is inversely proportional to the cognitive activity.

Table 2. Statistical validation using post hoc t-test to find significant differences in mean degrees

Networks	Participants	Mean difference	95% CI
NFC_Drive	P1	−15.2667*	[−17.6260, −12.9073]
vs.	P2	−17.9333*	[−20.0872, −15.7795]
PFC_Drive	P3	−17.0000*	[−19.6248, −14.3752]
	P4	−19.5333*	[−21.5160, −17.5506]
NFC_Drive	P1	4.1333*	[1.7740, 6.4927]
vs.	P2	2.2000*	[0.0462, 4.3538]
NFC_DriveAdo	P3	2.9333*	[0.3086, 5.5581]
	P4	2.5333*	[0.5506, 4.5160]
NFC_Drive	P1	−19.2667*	[−21.6260, −16.9073]
vs.	P2	−20.1333*	[−22.2872, −17.9795]
PFC_DriveAdo	P3	−19.9337*	[−22.5671, −17.3085]
	P4	−22.0667*	[−24.0494, −20.0840]
PFC_Drive	P1	19.4000*	[17.0407, 21.7593]
vs.	P2	20.1333*	[17.9795, 22.2872]
NFC_DriveAdo	P3	19.9333*	[22.5581, 17.3086]
	P4	22.0667*	[20.0840, 24.0494]
PFC_Drive	P1	−4.0000*	[−6.3593, −1.6407]
vs.	P2	−2.2000*	[−4.3538, −0.0462]
PFC_DriveAdo	P3	−2.9333*	[−5.5581, −0.3086]
	P4	−2.5333*	[−4.5160, −0.5506]
NFC_DriveAdo	P1	−23.4000*	[−25.7593, −21.0407]
vs.	P2	−22.3333*	[−24.4872, −20.1795]
PFC_DriveAdo	P3	−22.8667*	[−25.4914, −20.2419]
	P4	−24.6000*	[−26.5827, −22.6173]

*Mean difference is significant at α < 0.05 level

The post hoc t-test results with Bonferroni adjustment are presented in Table 2. The results show a substantial evidence against the null hypothesis that means of these populations are equal.

To further investigate the electrode sites that are hubs in the *NFC* and *PFC* networks of different cognitive load states, the degrees of them are plotted using topoplots as shown Figs. 6, and 7 for *Drive* and *DriveAdo* states respectively. An important observation from Fig. 6 is that many frontal electrode sites such as *FT7, F7, FP1, F8,*

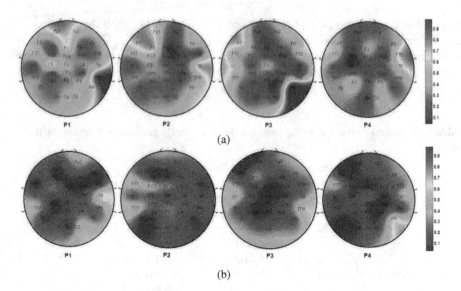

Fig. 6. Degrees of electrode sites of *NFC* networks (a) Drive (b) DriveAdo

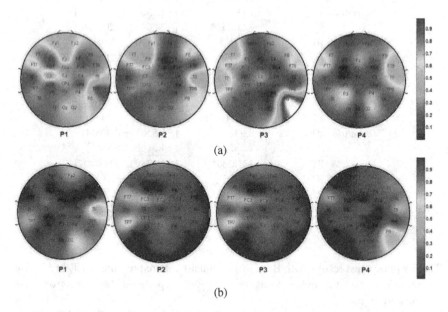

Fig. 7. Degrees of electrode sites of *PFC* networks (a) Drive (b) DriveAdo

FC3 and *FT8* and electrode sites *TP8, P8, T8* in temporal and parietal regions have many negative connections. On the other hand, during *DriveAdo*, the frontal region electrode sites have relatively a small number of negative connections. This analysis shows that frontal regions exhibit cohesive behavior with other brain regions during heavy cognitive load state.

From the results of various experiments, it is clear that the amount of negative correlation among the brain regions decreases with increase in cognitive load.

6 Conclusion

In this study, the impact of negative correlations present in the *FBNs* of mild and heavy cognitive load states is analyzed using graph theoretical approaches. Empirical analyses performed on *NFC* and *PFC* networks using connectivity density revealed that the number of negatively correlated edges decreases when the cognitive load increases. Multi-comparison test performed using the degrees of nodes in the *NFC* and *PFC* networks of *Drive,* and *DriveAdo* states revealed that the mean differences computed using the degrees of nodes in these networks are statistically significant. Moreover, the experimental analysis performed using the degrees of nodes in *NFC* and *PFC* networks showed that many electrode sites in the frontal region such as *FT7, F7, FP1, F8, FC3* and *FT8* have less number of negatively correlated edges during heavy cognitive load state (*DriveAdo*). In summary, all these empirical results show that the negative correlation between the brain regions decreases with increase in cognitive load. The *NFC* analysis has potential application in the diagnosis of cognitive impairments. Further analysis on *FBNs* with positive and negative correlations using signed graph analysis methods will help to analyze the stability of such networks.

Acknowledgement. This research work has been carried out in collaboration with the Cognitive Neuro-Engineering & Computational Neuroscience Laboratory (CNeL), University of South Australia, Australia.

References

1. Bullmore, E., Sporns, O.: Complex brain networks: graph theoretical analysis of structural and functional systems. Nat. Rev. Neurosci. **10**(3), 186–198 (2009)
2. Rubinov, M., Sporns, O.: Complex network measures of brain connectivity: uses and interpretations. Neuroimage **52**(3), 1059–1069 (2010)
3. Nunez, P.L.: Electroencephalography (EEG). In: Ramachandran, V.S. (ed.) Encyclopaedia of the Human Brain, pp. 169–179 (2002). editor in chief
4. Bressler, S.L., Menon, V.: Large-scale brain networks in cognition: emerging methods and principles. Trends Cogn. Sci. **14**, 277–290 (2010)
5. Cocks, B., Nandagopal, D., Vijayalakshmi, R., Thilaga, M., Dasari, N., Dahal, N.: Breaking the camel's back: can cognitive overload be quantified in the human brain? Procedia Soc. Behav. Sci. **97**, 21–29 (2013)
6. Nandagopal, D., et al.: Computational techniques for characterizing cognition using EEG data - new approaches. Procedia Comput. Sci. **22**, 699–708 (2013)
7. Sporns, O.: Structure and function of complex brain networks. Dialogues Clin. Neurosci. **15**, 247–262 (2013)

8. Stam, C.J., Reijneveld, J.C.: Graph theoretical analysis of complex networks in the brain. Nonlinear Biomed. Phys. **1**(1), 3 (2007)

9. De Vico Fallani, F., Richiardi, J., Chavez, M., Achard, S.: Graph analysis of functional brain networks: Practical issues in translational neuroscience. Philos. Trans. R. Soc. Lond. B **369**, 20130521 (2014)

10. Jalili, M.: Functional brain networks: does the choice of dependency estimator and binarization method matter? Sci. Rep. **6**, 29780 (2016)

11. Xu, T., et al.: Network analysis of functional brain connectivity in borderline personality disorder using resting-state fMRI. Neuroimage Clin. **11**, 302–315 (2016)

12. Chen, G., Chen, G., Xie, C., Li, S.J.: Negative functional connectivity and its dependence on the shortest path length of positive network in the resting-state human brain. Brain Connect. **1**(3), 195–206 (2011)

13. Kornbrot, D.: Pearson Product Moment Correlation. Encyclopedia of Statistics in Behavioral Science. Wiley, New York (2005). http://onlinelibrary.wiley.com. https://doi.org/10.1002/0470013192.bsa473

14. Buckner, R.L., et al.: Cortical hubs revealed by intrinsic functional connectivity: mapping, assessment of stability, and relation to Alzheimer's disease. Neuroscience **29**(6), 1860–1873 (2009)

15. Zhan, L., et al.: The significance of negative correlations in brain connectivity. J. Comp. Neurol. **525**(15), 3251–3265 (2017)

16. Vijayalakshmi, R., Dahal, N., Dasari, N., Cocks, B., Nandagopal, D.: Identification and analysis of functional brain networks. In: The Proceedings of International Conference on Pattern Recognition (ICPR) (2012)

17. Fröhlich, F.: Network Neuroscience, 1st edn. Academic Press, London (2016)

18. Wang, J.H., Zuo, X.N., Gohel, S., Milham, M.P., Biswal, B.B., He, Y.: Graph theoretical analysis of functional brain networks: test-retest evaluation on short- and long-term resting-state functional MRI data. PLoS ONE **6**(7), e21976 (2011). https://doi.org/10.1371/journal.pone.0021976

19. Chang, T.Y., et al.: Graph theoretical analysis of functional networks and its relationship to cognitive decline in patients with carotid stenosis. J. Cereb. Blood Flow Metab. **36**(4), 808–818 (2015)

20. Jacob, Y., et al.: Dependency network analysis (DEPNA) reveals context related influence of brain network nodes. Sci. Rep. **6**, 27444 (2016)

21. Zaslavsky, T.: Matrices in the theory of signed simple graphs. In: Proceedings of the International Conference on Discrete Mathematics, pp. 207–229 (2008)

22. Vijayalakshmi, R., Nandagopal, D., Dasari, N., Cocks, B., Dahal, N., Thilaga, M.: Minimum connected component - a novel approach to detection of cognitive load induced changes in functional brain networks. Neurocomputing **170**, 15–31 (2015)

23. Thilaga, M., et al.: A heuristic branch-and-bound based thresholding algorithm for unveiling cognitive activity from EEG data. Neurocomputing **170**, 32–46 (2015)

Frequent Sequence Mining Approach to Video Compression

M. Karthik[1]([✉]), C. Oswald[2], and B. Sivaselvan[2]

[1] Department of Information Technology, SSN College of Engineering, Chennai, India
karthik.murugesan@outlook.in
[2] Department of Computer Engineering, IIITDM Kancheepuram, Chennai, India
{coe13d003,sivaselvanb}@iiitdm.ac.in

Abstract. This work provides an approach of using frequent sequence mining in video compression. This paper focuses on reducing redundancies in a video by mining frequent sequences and then replacing it by the sequence identifiers. If we consider a video file as a sequence of raw RGB pixel values, we can observe a lot of redundancies and patterns/sequences that are repeated throughout the video. Redundant information and repeating sequences take up unnecessary space. The main motive of this system is to reduce these redundancies by employing data mining and coding techniques. The high cost of time and space required for mining sequences from large videos are reduced by dividing the video into multiple small blocks. Simulations of the proposed algorithm show a significant reduction in redundant parts of the video.

Keywords: Video compression · Frequent sequence mining
Lossy compression · Compression ratio

1   Introduction

Data transferred through the internet is increasing day by day and compression algorithms are necessary to make it more efficient and fast. Data exists in various forms like Text, Images, Audios, Videos. The transfer speed and the load on the network depend on the size of the file, thus the reduction in size is required as it is more expensive and time-consuming to upgrade the network hardware and capacity. Video compression is the process of reducing the memory needed to represent a video and is based on the fact pixel neighboring pixel values are correlated within a frame and other frames. Video compression based on pixels from a frame and its surrounding frames is known as spatial compression or inter-frame compression. If it is based on neighboring pixels in the same frame then it is known as temporal compression or inter-frame compression. Video compression techniques are broadly classified into lossless and lossy compression. Lossless compression reduces the size without reducing the quality of the video. Compression techniques like Shannon-Fano Encoding, Huffman Encoding, Arithmetic Encoding, Lempel-Ziv-Welch (LZW), Bit-Plane Coding

© Springer Nature Singapore Pte Ltd. 2018
G. Ganapathi et al. (Eds.): ICC3 2017, CCIS 844, pp. 87–97, 2018.
https://doi.org/10.1007/978-981-13-0716-4_8

and Lossless Predictive Coding perform lossless compression [6,11,12,15]. Lossy compression, on the other hand, reduced the size of the video by removing or reducing the irrelevancy in addition to redundancy, thereby achieving better compression than lossless compression. Compression techniques like H.264 [9], HEVC [14], MPEG-2 [5] perform the task of lossy video compression.

Data Mining is the process of extracting hidden and useful information from large DB's [4]. In Data Mining, five perspectives were observed by Ramakrishnan et al. which are Compression, Search, Induction, Approximation, and Querying [10]. The perspective of Data mining as a compression technique is of interest in this study. The process of data mining focuses on generating a reduced (smaller) set of patterns (knowledge) from the original DB, which can be viewed as a compression mechanism. A few of the mining techniques such as Classification, Clustering, Association Rule Mining (ARM) may be explored from the compression perspective. In this study, the knowledge Frequent Sequence Mining (FSM) is used to achieve efficient compression.

Frequent Sequence Mining (FSM) is the process of mining sequences that occur more frequently more than a minimum support count. FSM could be classified into two based on the representation of the DB, horizontal and vertical. An example of horizontal mining is GSP [13] which is similar to Apriori algorithm, and SPADE [16] and SPAM are examples of vertical mining algorithms.

Consider a set $I = \{a_1, a_2, a_3, \ldots, a_n\}$. I is considered as an itemset or n-itemset and elements a_i is considered as an item in itemset, where $0 \leq i \leq n$. A sequence s is represented as tuple, $I = \langle I_1 I_2 I_3 \ldots I_n \rangle$ where $I_i \subseteq I : 0 \leq i \leq n$. The number of items in the sequence in I is denoted by $|I|$, which is also the size of the sequence I. Consider the sequences $\alpha = \langle I_{a_1} I_{a_2} I_{a_3} \ldots I_{a_x} \rangle$ and $\beta = \langle I_{b_1} I_{b_2} I_{b_3} \ldots I_{b_y} \rangle$. We can say α is a sub sequence of β if there exists an $0 \leq i_1 \leq i_2 \leq i_1 \cdots \leq i_x \leq i_y$ such that $I_{a_1} \subseteq I_{b_{i_1}}, I_{a_2} \subseteq I_{b_{i_2}}, \ldots I_{a_x} \subseteq I_{b_{i_x}}$. Considering FSM-VC, a sequence is defined as a set of characters in an ordered and contiguous manner w.r.t the index. The words sequence and pattern are used interchangeably in this paper.

1.1   Related Work

In the area of lossy video compression, MPEG is one of the most common and widely used formats. Some of its versions that are widely used are MPEG-1, MPEG-2 and MPEG-4 part 10. MPEG-1, one of the early versions of MPEG is a standard for video lossy compression. This was design and created to compress analog video (VHS) to 1.5 Mbit/s with less quality loss, making it suitable for broadcasting through digital cable/satellite TV and digital audio broadcasting. MPEG-2 is an extension to MPEG-1 with a much higher bit rate, suitable for digital TV. MPEG-2 part 10 which is widely used in Youtube and Google video, compresses the video to almost half the size of that of MPEG-1 and MPEG-2 without compromising on the quality of the video. It is also known as H.264/AVC [7]. The major difference between H.264 and MPEG-1/2 is due to the support of arithmetic coding in addition to Huffman coding. There is also a support for variable size macroblocks from 4 by 4 to 16 by 16.

Video data is embedded into a container such as AVI or MP4. This is done mostly for reliable streaming of file with error correction. The compression method used in MPEG-1 and 2 is similar to JPEG image compression as it looks for similarities within the frame to reduce memory. But in addition to this compression is also done in between frames. A technique known as motion compensation is also used when looking in between frames. Considering the frames of a video, it can be categorized into 3 types, I (Inter-frame), P (predictive) and B (Bidirectional). These frames differ from others in the following ways. The I frame stands as the base for the other frame. It is not reduced in size much, but other frames are dependent upon the frames. A P frame is represented as how it's pixels differ from its previous frame. B frame is similar to P frames but also dependent upon the next frame. In video compression, motion compensation is used widely to reduce memory. It is a technique where movements in the frames could be used to predict the future frames. This is mostly applied to the P frames. If the encoder finds out that a part of the P frame is moving continuously, then this finding could be exploited by using the coding the moving P frame of its starting position and path and boundaries.

2   Proposed Architecture

2.1   Compressor

Source Video: The approach of FSM-VC is mentioned in Algorithm 1. The video V is chosen in the RGB color space and each component is converted to the form $m \times n \times f$, where m is the width, n is the height and f is the no. of frames in the video. Cell $\{i, j, k\}$ in this matrix represents the color value of the pixel in the ith column, j^{th} row, and k^{th} frame. The three color components of the video are represented by V_1, V_2 and V_3. The three components are independent of each other and are processed in parallel. Let's consider the matrix representation of each component V_i. The $m \times n \times f$ 3D matrix is converted to a 2D matrix with (m × n) rows and f columns. Each row in the matrix represents the values of each pixel for all the f frames of the video. Now we divide V_i column wise based on the block size b. The no. of blocks is represented using n_b and the blocks are represented as $B_i = \{B_{i_1}, B_{i_2}, \ldots, B_{i_{n_b}}\}$. The reason behind dividing it into multiple fixed size blocks is discussed in the coming sections. Each row in each block is considered as a transaction for mining. For example, consider a video of dimensions 320×240 and no. of frames f is 205, then the 2D matrix to which it is converted contains $320 \times 240 = 76,800$ rows and 205 columns. If we split this matrix into blocks of size $b = 100$, we get $n_b = 3$ blocks. The dimensions of these blocks will be 76800×100, 76800×100 and 76800×5. A sample trace is given below:

$$V_i = \begin{bmatrix} 2 & 12 & 133 & 24 & 25 & 7 & 11 & 92 & 10 & 113 & 13 & 44 & 15 \\ 5 & 12 & 3 & 14 & 15 & 7 & 128 & 9 & 102 & 61 & 13 & 143 & 15 \\ 5 & 12 & 3 & 1 & 1 & 7 & 218 & 19 & 12 & 61 & 10 & 123 & 25 \\ 5 & 12 & 3 & 4 & 15 & 7 & 217 & 7 & 100 & 68 & 10 & 144 & 15 \end{bmatrix}$$

Algorithm 1. *Compressor*

1: **Input :**
   1)Video file: V
   2)Block size: b
   3)Minimum support count: α
2: **Output :**
   1)Compressed file M
   2)Sequence table: ST
3: Split video V to V_1, V_2, V_3 corresponding to R, G, B components of the video, and each component is of the dimensions m × n × f
4: Number of blocks $n_b = \dfrac{f}{b}$
5: $i \leftarrow 1$
6: **for** i in [**1, 3**] **do**
7:     Split V_i into n_b blocks $B_{i_1}, B_{i_2}, \ldots, B_{i_{n_b}}$ of block size $\leq b$
8:     $[S_{i_1}, S_{i_2}, ..., S_{i_{n_b}}] = CM\_SPAM([B_{i_1}, B_{i_2}, ..., B_{i_{n_b}}], \alpha)$
9:     $[S'_{i_1}, S'_{i_2}, ..., S'_{i_{n_b}}] = Coding([S_{i_1}, S_{i_2}, ..., S_{i_{n_b}}])$
10:    $[B'_{i_1}, ..., B'_{i_{n_b}}], [D_{i_1}, ..., D_{i_{n_b}}] = Replace([S'_{i_1}, S'_{i_2}, ..., S'_{i_{n_b}}], [B_{i_1}, B_{i_2}, ..., B_{i_{n_b}}])$
11:    $M_i = Combine([B'_{i_1}, B'_{i_2} ..., B'_{i_{n_b}}], [D_{i_1}, D_{i_2} ..., D_{i_{n_b}}])$
12:    $ST_i = [S'_{i_1}, S'_{i_2}, ..., S'_{i_{n_b}}]$
13: **end for**
14: $M = Merge(M_1, M_2, M_3)$
15: **return** M, ST

$$B_{i_1} = \begin{bmatrix} 2 & 12 & 133 & 24 & 25 \\ 5 & 12 & 3 & 14 & 15 \\ 5 & 12 & 3 & 1 & 1 \\ 5 & 12 & 3 & 4 & 15 \end{bmatrix} \quad B_{i_2} = \begin{bmatrix} 7 & 11 & 92 & 10 & 113 \\ 7 & 128 & 9 & 102 & 61 \\ 7 & 218 & 19 & 12 & 61 \\ 7 & 217 & 7 & 100 & 68 \end{bmatrix} \quad B_{i_3} = \begin{bmatrix} 13 & 44 & 15 \\ 13 & 143 & 15 \\ 10 & 123 & 25 \\ 10 & 144 & 15 \end{bmatrix}$$

In the above example V_i is the 2D matrix representation of a component of a sample video. It has $m * n = 4$ rows and $f = 15$ columns. If we divide V_i into $n_b = 3$ blocks column wise, we get the blocks B_{i_1}, B_{i_2} and B_{i_3} as given above.

Mining: The blocks B_i mentioned in the previous section is the input for the mining operation. Each row in B_i s considered as a single transaction and frequent sequences are mined using **CM-SPAM** [3]. CM-SPAM uses the vertical representation of the database to mine frequent sequences making it efficient in long transactions. The pseudo code of CM-SPAM is given in Algorithm 5 take as input a sequence database SDB and the *minsup* threshold. SPAM first constructs the vertical representation V(SDB) from the input database SDB. After the vertical representation is constructed, set of frequent itemsets with only a single element F_1 is found. Then for each element s in F_1, the SEARCH procedure is called from SPAM with parameters $\langle s \rangle$ and F_1 where support count of s is larger than *minsup*. The support count or support of a set s is an indication of how frequently the elements of the set appear in the DB. The SEARCH procedure recursively explores pattern with the prefix as $\langle \{s\} \rangle$. In each cycle,

SEARCH generate two types of patterns, the s-extension and the i-extension. The s-extension of the sequence $\langle P_1, P_2, \ldots, P_h \rangle$ with an item s is given by $\langle P_1, P_2, \ldots, P_h, \{s\} \rangle$. The i-extension of the sequence $\langle P_1, P_2, \ldots, P_h \rangle$ with item s is given as $\langle P_1, P_2, \ldots, P_h \cup \{s\} \rangle$. For each sequence generated using the extension operation the support count is determined. Sequences with support count less then $minsup$ are pruned by not extending those sequences as they are infrequent. This is based on the property which states that a frequent sequence could not be obtained from a infrequent sequence [1]. Bitmap representation [2] of the vertical database is used for calculating support count faster. In this paper, this algorithm is used to mine frequent sequences from each block B_i separately. Each block is represented in such a way that each row is a sequence and each item set in a sequence consists of only one element. Due to this, there is no need for generating i-extension at each level. An example is given below.

SDB		1		2		3	
SID	Sequences	SID	Itemsets	SID	Itemsets	SID	Itemsets
1	$\langle \{1\}, \{2\}, \{3\}, \{1\} \rangle$	1	1, 4	1	2	1	3
2	$\langle \{2\}, \{3\}, \{2\}, \{1\} \rangle$	2	4	2	1, 3	2	2
3	$\langle \{2\}, \{2\}, \{3\}, \{1\} \rangle$	3	4	3	1, 2	3	3

The vertical databases of SDB is given in Tables 1, 2 and 3. F_1, F_2, F_3 and F_4 for SDB is given below:

F_1		
PID	Seq.	Supp.
1	$\langle \{1\} \rangle$	3
2	$\langle \{2\} \rangle$	3
3	$\langle \{3\} \rangle$	3

F_2		
PID	Seq.	Supp.
1	$\langle \{1\}, \{1\} \rangle$	0
2	$\langle \{1\}, \{2\} \rangle$	1
3	$\langle \{1\}, \{3\} \rangle$	0
4	$\langle \{2\}, \{1\} \rangle$	1
5	$\langle \{2\}, \{2\} \rangle$	1
6	$\langle \{2\}, \{3\} \rangle$	2
7	$\langle \{3\}, \{1\} \rangle$	2
8	$\langle \{3\}, \{2\} \rangle$	1
9	$\langle \{3\}, \{3\} \rangle$	0

F_3		
PID	Seq.	Supp.
1	$\langle \{2\}, \{3\}, \{1\} \rangle$	2
2	$\langle \{2\}, \{3\}, \{2\} \rangle$	1
3	$\langle \{2\}, \{3\}, \{3\} \rangle$	0
4	$\langle \{3\}, \{1\}, \{2\} \rangle$	0
5	$\langle \{3\}, \{1\}, \{2\} \rangle$	0
6	$\langle \{3\}, \{1\}, \{3\} \rangle$	0

F_4		
PID	Seq.	Supp.
1	$\langle \{2\}, \{3\}, \{1\}, \{1\} \rangle$	0
2	$\langle \{2\}, \{3\}, \{1\}, \{2\} \rangle$	0
3	$\langle \{2\}, \{3\}, \{1\}, \{3\} \rangle$	0

PID	Seq.	Supp.
1	$\langle \{2\}, \{3\} \rangle$	2
2	$\langle \{3\}, \{1\} \rangle$	2
3	$\langle \{2\}, \{3\}, \{1\} \rangle$	2

The maximum length of the frequent sequences generated in the mining process mentioned above depends upon the maximum length of the transactions, as there exists no frequent sequence larger than the transaction itself. Thus dividing the video into multiple fixed size blocks limits the maximum length of the frequent sequences mined from the block. We can observe in each iteration of CM-SPAM recursively generates multiple sequences by extension operations.

Now limiting the size of the transactions limits the size of the recursive tree generated, by which the memory requirement could be controlled. This is the main reason behind splitting the data into multiple fixed sized blocks.

Algorithm 2. *Coding*

1: **Input:** Mined sequences $S = [S_{i_1}, S_{i_2}, ..., S_{i_{n_b}}]$
2: **Output:** Modified Sequences $[S'_{i_1}, S'_{i_2}, ..., S'_{i_{n_b}}]$
3: **for** j in $[1, n_b]$ **do**
4:      $size$ = length of S_{i_j}
5:      $[sup_1, sup_2, ..., sup_{size}]$ = support count of each sequence in S_{i_j}
6:      $[len_1, len_2, ..., len_{size}]$ = support count of each sequence in S_{i_j}
7:      $[er_1, er_2, ..., er_{size}] = [sup_1 \times len_1, sup_2 \times len_2, ..., sup_{size} \times len_{size}]$s
8:      S'_{i_j} = sort(S_{i_j}) in descending order of $[er_1, er_2, ..., er_{size}]$
9:      $S'_i = [S'_{i_1}, S'_{i_2}, ..., S'_{i_{n_b}}]$
10: **end for**
11: **Return** S'_i

Coding: The frequent sequences S_i is obtained from the mining step along with the support count sup for each sequence. Each sequence has its own length len_i. Now we compute S'_i by finding $er = len \times sup$ for each sequence in S_i and sorting the sequences in the descending order of er. We define er as the effective no. of pixels the code of the particular sequence will be replacing. Now we assign a unique code for each sequence in S'_i. Each sequence is given a code corresponding to the index value in the list of sequences. For example, the 5th sequence in S'_i will be represented by the code '5'. The pseudo code of this process is given in Algorithm 2. An example trace is given below:

$$S_{i_k} = \begin{bmatrix} 1\ 5\ 3\ 2\ 1 \\ 1\ 2\ 3\ 4\ 7\ 8 \\ 4\ 6\ 7 \end{bmatrix} \quad len = \begin{bmatrix} 5 \\ 6 \\ 3 \end{bmatrix} \quad sup = \begin{bmatrix} 124 \\ 528 \\ 1002 \end{bmatrix} \quad len \times sup = \begin{bmatrix} 620 \\ 3168 \\ 3006 \end{bmatrix}$$

$$S'_{i_k} = \begin{bmatrix} 1\ 2\ 3\ 4\ 7\ 8 \\ 4\ 6\ 7 \\ 1\ 5\ 3\ 2\ 1 \end{bmatrix} \quad code = \begin{bmatrix} 1 \\ 2 \\ 3 \end{bmatrix} \quad er = \begin{bmatrix} 3168 \\ 3006 \\ 620 \end{bmatrix}$$

In the above example, the mined sequences for block B_{i_k} is represented by S_{i_k}. The length len and the support count sup of each sequence is also given. er is calculated as $len \times sup$. S'_{i_k} is created by sorting S_{i_k} according to it's corresponding er in the descending order. After obtaining S'_{i_k}, we assign codes to each sequence in a sequential order as represented by the matrix *code*.

Replace: We use B_i and S'_i obtained from the *Coding* part to perform the *Replace* operation. The pseudo code is given in Algorithm 3. This operation replaces the sequences S'_{i_k} obtained for block B_{i_k} with its corresponding code and add the index where the code was replaced to the dictionary corresponding to the row. Consider the following example:

$$r = \begin{bmatrix} 1\ 5\ 3\ 2\ 5\ 6 \end{bmatrix} \quad s_4 = \begin{bmatrix} 1\ 5\ 3\ 2 \end{bmatrix} \quad r' = \begin{bmatrix} 4\ 5\ 6 \end{bmatrix} \quad d = \begin{bmatrix} 1 \end{bmatrix}$$

Here consider a row/transaction r in a block and a sequence s represented by code '4'. After replacing the occurrence of s in r by its corresponding code, we get r'. The index at which the code was replaces is stored in the dictionary D. Each row will have its own dictionary to indicate the indices where sequence codes's where replaced.

Algorithm 3. *Replace*

Input:
1)$S_i' = [S_{i_1}', S_{i_2}', ..., S_{i_{n_b}}']$
2)$B_i = [B_{i_1}, B_{i_2}, ..., B_{i_{n_b}}]$
Output:
1) $B_i' = [B_{i_1}', B_{i_2}', ..., B_{i_{n_b}}']$
2) $D_i = [D_{i_1}, ..., D_{i_{n_b}}]$
for $j \in [1, n_b]$ do
    $D_{i_j} = \{\ \}$
    for p in $[1, \text{length of } S_{i_j}]$ do
        for k in B_{i_j} do
            if $S_{i_j}[p]$ is in k then
                Replace $S_{i_j}[p]$ in k
                Add p in D_{i_j}
            end if
        end for
    end for
end for
return $[B_{i_1}, B_{i_2}, \ldots, B_{i_{n_b}}], [D_{i_1}, D_{i_2}, \ldots, D_{i_{n_b}}]$

Algorithm 4. *Combine*

1: **Input :**
    1) $B_i' = [B_{i_1}', B_{i_2}' ..., B_{i_{n_b}}']$
    2) $D_i = [D_{i_1}, D_{i_2} ..., D_{i_{n_b}}]$
2: **Output :** M_i
3: **for** j in $[1, n_b]$ **do**
4:     $BD_i = $ Append each row of B_{i_j} with the respective rows of D_{i_j}
5: **end for**$M_i = \{\ \}$
6: **for** j in $[1, n_b]$ **do**
7:     Row wise append BD_{i_j} to M_i
8: **end for**
9: **return** M_i

Combine: The pseudo code for the combine operation is given in Algorithm 4. In this function each block B'_{i_k} in B'_i is combined with each other. The blocks are separated with a separating bit to identify the split. The dictionary D_i is combined in a similar way as B'_i.

Algorithm 5. SPAM$(SDB, minsup)$

1: Scan SDB to create V(SDB) and identify F_1, the list of frequent items.
2: **for** each item s $\in F_1$ **do**
3:    $Search(\langle s \rangle, F_1, \{e \in F_1 | e >_{lex} s\}, minsup)$
4: **end for**

Algorithm 6. Search$(pat, S_n, I_n, minsup)$

1: Output pattern pat
2: $S_{temp} := I_{temp} := \emptyset$
3: **for** each item j in S_n **do**
4:    **if** the s-extension of pat is frequent **then**
5:       $S_{temp} := S_{temp} \cup \{i\}$
6:    **end if**
7: **end for**
8: **for** each item j in S_{temp} **do**
9:    $Search$(the s-extension of pat with j, S_{temp}, $\{e \in S_{temp} \mid e >_{lex} j\}$, minsup)
10: **end for**
11: **for** each item j in I_n **do**
12:    **if** the i-extension of pat is frequent **then**
13:       $I_{temp} := I_{temp} \cup \{i\}$
14:    **end if**
15: **end for**
16: **for** each item j in I_{temp} **do**
17:    $Search$(the i-extension of pat with j, I_{temp}, $\{e \in I_{temp} \mid e >_{lex} j\}$, minsup)
18: **end for**

---

3 Simulation Results and Discussion

The simulation is performed on an Intel core i5 CPU, 1.6 GHz, 8 GB Main Memory, 128 GB Secondary Flash memory and the algorithm is implemented in Python3. This algorithm is tested on various videos such as drone, parking and bunny of 320×280 in avi format taken from the VIRAT [8] video dataset. At the moment the algorithm has only been tested in videos of smaller frame size and memory to test the feasibility of the algorithm. Larger videos will be tested in the future (Table 1).

Talking about the videos datasets, each video is different from each other and was chosen to test certain aspects such as compressing stable areas, handling random movements by the camera and sudden changes of views. The videos are bunny, drone, and parking. The bunny video consists of a few scenes with

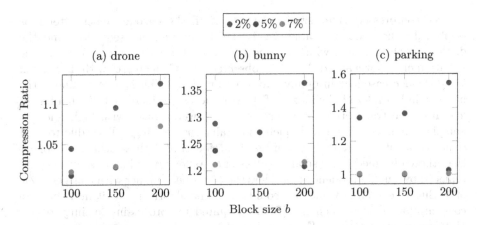

Fig. 1. Compression Ratio C_r vs the Block size b for drone, bunny and parking

Table 1. Values of various parameters for different video samples at $\alpha = 2\%$

Video	Original size (Mb)	b	Compressed file size (Mb)	Sequence table size (Mb)	C_r	H.264 C_r	MPEG-2 C_r
Bunny	1.1	100	0.750	0.104	1.289	1.416	1.436
		150	0.758	0.107	1.272		
		200	0.776	0.135	1.207		
Drone	4.4	100	4.2	0.011	1.045	1.57	1.76
		150	4	0.016	1.096		
		200	3.9	0.01	1.125		
Parking	3.5	100	2.6	0.011	1.340	23.1	1.842
		150	2.5	0.063	1.366		
		200	2.2	0.060	1.549		

certain areas stable for each scene. Sudden changes to the entire frame between the scenes in bunny can be seen. In drone, the whole video is inconsistent and there is no fixed position of the camera. A lot of rapid and fast changes to the view of the camera are present and less no. of stable pixels could be observed in the video. Considering the parking video, it is shot from a CCTV camera in a parking lot, and it has a large no. of stable pixels across the frame and very few moving pixels could be observed.

The parameters b(block size) and α(minimum support) affect the compression ratio of the video. Hence the results are observed by varying the parameters b and α to study their effects on compression. The compression ratio C_r is defined as

$$C_r = \frac{\text{Uncompressed size of video}}{\text{Compressed size of video} + \text{Sequence table size}}$$

The compressed video size is the size of all the encoded block after it is combined. The Sequence table size is the list of frequent sequences and the dictionary D combined which is used to encode the video. Figure 1 shows the variation of C_r for b and α. It can be observed that C_r increases with increase in b and the decrease in α in bunny, drone, and parking. This is because due to the increase in block size b the length of the transactions is increased. With a higher length of the transactions, frequent sequences were mined with higher length and the number of frequent sequences obtained is also high. Thus due to this, sequence identifiers represent more sequences and also those which are longer than those obtained using smaller values of b. As longer and more sequences are replaced by smaller sequence identifiers the C_r value is higher with a larger value of b. But a very large value of b could affect the algorithms as it may increase consumption of main memory (RAM) required for processing leading to Out of Memory exceptions. A Higher value of b produces a large recursive tree in the CM-SPAM mining operation. If the resulting recursive tree is large, Out of Memory exceptions occur. Considering the increase of C_r with decrease in α, it is because with lower value of α more number of sequences are mined from the blocks than those obtained with higher values of α. Thus more number of sequences are replaced with small sequence identifiers compared to the higher values of α. Due to this, a lower α value produces a higher C_r. Consider the datasets the algorithm was tested on. In parking video represented by Fig. 1a, we observe a highest value of C_r is observed with $\alpha = 2$ and $b = 200$. This is because the video is composed of only one camera view and there is very less moving parts in the video. In drone represented by 1b, due to the lack of a fixed view and a lot of movement, longer patterns could be rarely found and thus a good C_r couldn't be achieved as t. Considering bunny represented by 1c, due to the presence of a few stable views and an average number of moving pixels, sequences were mined with moderate length and thus it was able to achieve a much better C_r than drone.

Observing the above results, FSM-VC was able to punish videos with a large number of stable pixels and fixed camera view, but performs poorly on videos with a lot of random movements and unstable pixels. Thus it could be used in videos with a large number of stable areas with the same color values to reduce a lot of storage memory. This algorithm presents a simple and basic approach where sequence mining could be applied to video compression.

4   Conclusion

We presented a novel, simple and basic sequence mining based lossy video compression algorithm for videos. We have proved that our method is good for mining frequent sequences and is most of the time comparable to the other compression algorithms. Even then we consider our algorithm as a prototype that needs further improvements. We intend to equip our algorithm with a fast and less resource consuming mining technique that would provide a significant improvement to the compression ratio and the time and memory required. This algorithm could be used as a base in development of an even more complex algorithms. Addition to that we would like to investigate out algorithm with larger video datasets and to explore more parameters that would affect the performance.

References

1. Agrawal, R., Srikant, R.: Mining sequential patterns. In: Proceedings of the Eleventh International Conference on Data Engineering, pp. 3–14. IEEE (1995)
2. Ayres, J., Flannick, J., Gehrke, J., Yiu, T.: Sequential pattern mining using a bitmap representation. In: Proceedings of the Eighth ACM SIGKDD International Conference on Knowledge Discovery and Data Mining, pp. 429–435. ACM (2002)
3. Fournier-Viger, P., Gomariz, A., Campos, M., Thomas, R.: Fast vertical mining of sequential patterns using co-occurrence information. In: Tseng, V.S., Ho, T.B., Zhou, Z.-H., Chen, A.L.P., Kao, H.-Y. (eds.) PAKDD 2014. LNCS (LNAI), vol. 8443, pp. 40–52. Springer, Cham (2014). https://doi.org/10.1007/978-3-319-06608-0_4
4. Han, J., Pei, J., Kamber, M.: Data Mining: Concepts and Techniques. Elsevier (2011)
5. Haskell, B.G., Puri, A., Netravali, A.N.: Digital Video: An Introduction to MPEG-2. Springer, Dordrecht (1996). https://doi.org/10.1007/b115887
6. Huffman, D.A.: A method for the construction of minimum-redundancy codes. Proc. IRE $40(9)$, 1098–1101 (1952)
7. Marpe, D., Wiegand, T., Sullivan, G.J.: The H.264/MPEG4 advanced video coding standard and its applications. IEEE Commun. Mag. $44(8)$, 134–143 (2006)
8. Oh, S., Hoogs, A., Perera, A., Cuntoor, N., Chen, C.C., Lee, J.T., Mukherjee, S., Aggarwal, J., Lee, H., Davis, L., et al.: A large-scale benchmark dataset for event recognition in surveillance video. In: 2011 IEEE Conference on Computer Vision and Pattern Recognition (CVPR), pp. 3153–3160. IEEE (2011)
9. Ostermann, J., Bormans, J., List, P., Marpe, D., Narroschke, M., Pereira, F., Stockhammer, T., Wedi, T.: Video coding with H.264/AVC: tools, performance, and complexity. IEEE Circ. Syst. Mag. $4(1)$, 7–28 (2004)
10. Ramakrishnan, N., Grama, A.Y.: Data mining: from serendipity to science. Computer $32(8)$, 34–37 (1999)
11. Salomon, D.: Data Compression: The Complete Reference. Springer, London (2004). https://doi.org/10.1007/978-1-84628-603-2
12. Shannon, C.E.: Communication theory of secrecy systems. Bell Labs Tech. J. $28(4)$, 656–715 (1949)
13. Srikant, R., Agrawal, R.: Mining sequential patterns: generalizations and performance improvements. In: Apers, P., Bouzeghoub, M., Gardarin, G. (eds.) EDBT 1996. LNCS, vol. 1057, pp. 1–17. Springer, Heidelberg (1996). https://doi.org/10.1007/BFb0014140
14. Sullivan, G.J., Ohm, J., Han, W.J., Wiegand, T.: Overview of the high efficiency video coding (HEVC) standard. IEEE Trans. Circ. Syst. Video Technol. $22(12)$, 1649–1668 (2012)
15. Witten, I.H., Neal, R.M., Cleary, J.G.: Arithmetic coding for data compression. Commun. ACM $30(6)$, 520–540 (1987)
16. Zaki, M.J.: Spade: an efficient algorithm for mining frequent sequences. Mach. Learn. $42(1)$, 31–60 (2001)

Deep Learning Based Audio Scene Classification

E. Sophiya[1(✉)] and S. Jothilakshmi[2]

[1] Department of Computer Science and Engineering, Annamalai University,
Annamalainagar, India
venus.sophiya@gmail.com
[2] Department of Information Technology, Annamalai University,
Annamalainagar, India
jothi.sekar@gmail.com

Abstract. Acoustic Scene Classification (ASC) is defined as recognition and categorizing an audio signal that identifies the environment in which it has been produced. This work aims to develop a Deep Neural Network (DNN) based system to detect the real life environments by analyzing their sound data. Log Mel band features are used to represent the characteristics of the input audio scenes. The parameters of the DNN are set according to the DNN baseline of DCASE 2017 challenge. The system is evaluated with TUT dataset (2017) and the result is compared with the baseline provided. The evaluation of proposed model shows an accuracy of 82%, which is better than the baseline system.

Keywords: Audio processing · Audio scene analysis
Audio scene classification · Deep learning · Audio features

1 Introduction

An audio scene is a blend of background noise and variety of foreground sound events. In comparison to speech, Environmental sounds are distinct and extent to a large scale of applications. For human it is an easy task because our brain performs these complex calculations and provides us wide ranging experience. This makes human to easily identify and separate various sounds within an auditory scene. But this is a complex task for an artificial system. Machine learning systems have more difficulties in finding the perception of human auditory systems in realistic acoustic scenes. This problem may seem to be challenging due to huge variety of sound sources in everyday environment. Since recognizing various types of environmental sound may provide potential usages, researches are focused to the growing field of Auditory Scene Analysis (ASA) [1]. This field of study is also known as Computational Auditory Scene Analysis (CASA). The concept was first defined by Bregman [2] and he states that the goal of ASA is to produce separate streams from the auditory input, such that each stream represents a single source in the acoustic environment.

G. Ganapathi et al. (Eds.): ICC3 2017, CCIS 844, pp. 98–109, 2018.
https://doi.org/10.1007/978-981-13-0716-4_9

The field of ASA focuses on Acoustic Scene Classification (ASC) and Acoustic Event Detection (AED) [3, 4]. An ASC system aims to classify the environment in which the scene was recorded (e.g. busy street, office, metro station) while an AED system focuses on sound events present within an audio [5, 6]. ASC and AED have applications [5, 7] in audio classifications [8], security surveillance [9, 10], military and public abnormal event detection [11], audio indexing [12–14], and ambient assistive living [15]. ASC is defined as recognizing and associating an acceptable label to an audio stream that identifies the environment in which it has been produced [4]. These audio scenes can be defined according to various geographical backgrounds (beach, park, busy roads, and streets), various indoor or outdoor areas (cafes, office, home, market, and library) and other transport backgrounds (car, bus, train, subways) [16]. The objective of ASC is to recognize the environment using acoustic signals. ASC is one of the important challenges in the field of CASA. The computational algorithm analyzes the sounds and identifies the environment using signal processing and machine learning mechanism (Fig. 1).

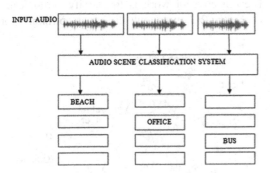

Fig. 1. Audio scene classification

The existing systems for ASA have widely used the traditional classifiers such as Support Vector Machine (SVM), Gaussian Mixture Model (GMM) which do not have the feature abstraction capability found in deeper models. Recently, deep learning is becoming more popular for supervised machine learning applications such as environmental sound classification, robust audio event detection, and speech recognition.

A deep learning based system for audio scene classification is proposed in this work. Multi label feed-forward Deep Neural Network (DNN) architecture to various feature representations are generated from signal processing methods. The Mel Frequency Cepstral Coefficients are extracted from short time frames of audio signal. These features are used to train multi-class DNN for learning decision function.

The rest of the paper is described as follows. The proposed DNN framework and classification approach is presented in Sect. 2. Experimental results obtained by the proposed approach are discussed in Sect. 3. Finally, Sect. 4 concludes the paper.

2   Proposed Audio Scene Classification System (DNN)

A DNN is an Artificial Neural Network (ANN) modelled with inspiration of human neural networks to recognize the patterns in real world data [17]. The traditional neural network consists of input layer, output layer and at most one hidden layer. The increase of hidden layers in the network makes it deep learning network. DNN can be used for supervised or unsupervised learning based on feature representation [18]. The reason for the popularity of deep learning is the availability of large volume of data, exponential growth of computational power, advancement in scaling of algorithms.

2.1   Network Architecture

DNN is a non-linear multilayer perceptron model with a powerful capability to extract relevant features related to classification [5, 6]. The multilayer perceptron (MLP), also called Feed Forward Network, is the most typical neural network model. Learning the feature representations automatically from the data is the major advantage of deep learning network. The objective of work is to classify audio feature vector among acoustic scene classes. Figure 2 shows a multi-layer perceptron network of three layers.

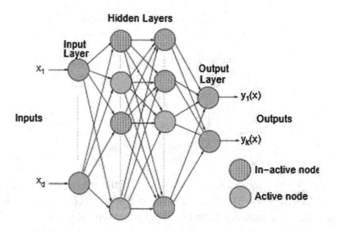

Fig. 2. Deep neural networks architecture

The DNN used in this work is fully connected neural network which consists of one input layer, one output layer, and several hidden layers. The number of neurons of input layer depends on the dimensions of input feature vector, while the number of neurons in output layer should be equal to the number of acoustic scenes being considered [19].

2.2 Feature Extraction

Audio features act as a quantitative way to provide the most important information in an audio file. The process of extracting relevant characteristics enclosed within the input data is called as feature extraction. This process converts an audio signal into a sequence of feature vectors. Feature extraction reduces the redundant information from audio signal and provides a compact representation.

Some of audio features are Temporal Domain Features, Frequency Domain (physical), Frequency Domain (perceptual), Cepstral Domain [20]. There are many feature extraction techniques such as Linear Predictive Analysis (LPC), Zero Crossing Rate (ZCR), Linear Predictive Cepstral Coefficients (LPCC), Perceptual Linear Predictive Coefficients (PLP), Mel-Frequency Cepstral Coefficients (MFCC), Power spectral Analysis (FFT), Mel scale Cepstral analysis (MEL), Relative spectra filtering of the log domain coefficients (RASTA).

The most common features used in speech recognition [5, 19, 21] are the MFCCs, Mel-band energy features, Mel-spectrogram. These are used with traditional classifiers such as Gaussian Mixture Model (GMM), Support Vector Machine (SVM), and Hidden Markov Model (HMM). MFCC is a representation of short term power spectrum of a sound. To compute MFCC, initially the audio signal is divided into short frames of 40 ms with an overlap of 50% of frame size. Then each short time frame is multiplied with a hamming window to maintain the continuity of audio signal in the short frames. The Discrete Fourier Transform is applied on each short frame. The next step is to take magnitude coefficients and multiply with the filter gain. Since computation of inverse Fourier Transform is expensive, Discrete Cosine Transform (DCT) is applied to extract MFCC. The steps involved in the computation of MFCC are illustrated in Fig. 3.

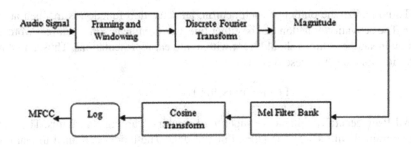

Fig. 3. MFCC feature extraction

2.3 Classification

After the feature selection process it is important to classify the input audio signal. Classification is the process by which a particular label is assigned to a particular audio format. The label would define the signal and its origin. A classifier defines decision boundaries in the feature space, which separates different sample classes from each other.

Multi-layer Perceptron consists of multiple layers interconnected to form a feed-forward neural network. In this network each neuron in one layer is directly connected to neurons in next layer. The input layer receives the input feature vector and the output layer predicts the decision of input. The hidden layers in between the input and output layer can be more in number. At each layer the neuron will be fired by calculating a net value α with weighted sum of input x_i and w_i respectively along with a bias b. The basic unit in this model is a neuron which is shown in Fig. 4.

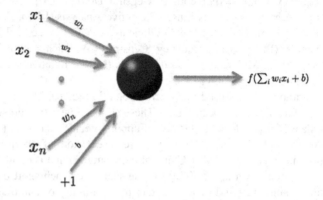

Fig. 4. Basic unit of neuron

α is given by

$$\alpha = \sum w_i x_i + b \tag{1}$$

To normalize the output of each neuron, an activation function is mapped at each layer. The activation function can be either linear or nonlinear. Since our real word data are nonlinear, the neurons should learn with nonlinear representations. Thus a nonlinear activation function f is described as

$$\mathbf{f}(\alpha) = \mathbf{max}(\mathbf{0}, \alpha) \qquad \mathbf{f}(\cdot)\varepsilon\mathbf{R}_+ \tag{2}$$

MLP are generally suitable for supervised machine learning algorithms. The system will be trained with a set of input and output data. The model is trained to learn with these training data. While training the model, parameters such as weights and bias can be adjusted in order to minimize the error rate including root mean square error (RMSE). In output layer the classification will be predicted by using a softmax loss classifier. Deep learning is started using random initialization based on Normal or Uniform distribution to specify scaling parameters.

2.4 Regularization

The goal of machine learning algorithm is to make the model to learn from training data and to predict an unseen future data. A common problem in machine learning is getting over fitting, which occurs due to noise in data and having large number of parameters than training data. This makes the model to get complex with low training error and high test error. The deep learning supports regularization techniques to avoid over fitting within the model [22]. The regularization is sum of squared coefficients of the model multiplied by a lambda function. L1 (Lasso) and L2 (Ridge) regularization are used to minimize the loss function. L1 regularization, represents the sum of all L1 norms of the weights and biases by shrinking most of the coefficients to zero. L2 regularization, represents the sum of squares of all the coefficients. The constants 1 and 2 are generally very small [23].

The deep learning model uses one more powerful regularization technique called dropout. Dropout is a technique which randomly ignores few neurons during training which provides model optimization. The neurons will be randomly dropped at each layer based on probability. This reduces the model getting over fitted and improves the performance of supervised learning. Dropout can be activated at each layer among either one of TanhWithDropout, RectifierWithDropout and MaxoutWithDroput. The probability of neurons to be dropped in each layer can be specified by dropout ratios.

3 Experiments and Results

In this system, the official dataset of the IEEE AASP Challenge (2017) on Detection and Classification of Acoustic Scenes and Events (DCASE) is used. This dataset consists of 15 acoustic scenes which are Beach, Bus, Cafe, Car, City center, Forest path, Grocery store, Home, Library, Metro, Office, Park, Resident, Train, and Tram. Each recording contains 10 s segments. The audio segments are in wave file format with 44.1 kHz and 16 bit stereo. For this work, the audio segments are converted to 16 bit mono. But the original sampling frequency is retained without down sampling since meaningful spectral characteristics will be found in high frequency range. The implementation of deep learning architecture uses log Mel band energy features. The features are calculated using 40 ms frames with a 50% overlap. The feature vector is constructed using 5 frame context which results in feature vector of length 200.

For classification of audio scenes, the dataset is partitioned randomly into three parts: 60% for Training, 20% for Validation, 20% for final Testing. The proposed model is evaluated using fourfold cross validation scheme. For each fold, per-class accuracy is calculated on a frame-wise level. These scores are obtained by dividing number of correctly predicted frames by the total number frames belonging to that class. Finally, the overall score is calculated by averaging the cross fold accuracy.

The parameters of the DNN are set according to the DNN baseline of DCASE 2017 challenge. The traditional activation function used in each hidden layer is replaced by Rectified Linear Units (ReLUs). The function ReLUs [5, 6, 22] has faster computation than sigmoid function and it is applied to all layers except output layer which uses Softmax output with a categorical cross-entropy loss function. DNN has a problem of over fitting with small training data. Dropout technique has been used to avoid this problem [22]. The learning rate is set to 0.001 and initially the number of epochs is set to 10. The performance of DNN is evaluated by varying the number of epochs upto 200. The momentum is set to 0.2. The input dropout ratio is initialized as 0.2 and for both hidden layers it is 0.2. Initial weight distribution is considered as normal distribution.

For this work, R machine learning library 'H2O Deep learning' is used to implement DNN. H2O is an open-source machine learning library for deep learning Applications which highly supports large scale data. Advanced algorithms, like Deep Learning, Boosting, and Bagging Ensembles are built-in to help application designers create smarter applications through elegant Application Program Interfaces (API).

The results obtained from the proposed system using the development mode cross validation setup is shown below. The baseline system is multi-layer perceptron architecture with 40 log Mel band energy features. Using these features, a neural network with two dense layers of 50 hidden units per layer is trained for 200 epochs. The classification decision is based on the accuracy metrics. The averaged overall evaluation folds result of baseline system is 74% of accuracy. Acoustic Scene Classification results averaged over each class and compared with baseline system is shown in Table 1.

Table 1. Comparison of class-wise accuracy with baseline system.

Acoustic scene	Baseline system	Proposed system
Beach	75.3%	88%
Bus	71.8%	74%
Café	57.7%	75%
Car	97.1%	85%
City center	90.7%	91%
Forest path	79.5%	90%
Grocery store	58.7%	76%
Home	68.6%	79%
Library	57.1%	70%
Metro	91.7%	85%
Office	99.7%	92%
Park	70.2%	78%
Resident	64.1%	67%
Train	58.0%	71%
Tram	81.7%	67%

The proposed model achieves better performance when compared to the baseline system with given parameters. The overall confusion matrix is shown in Fig. 5.

The cross fold evaluation of proposed system is shown in Table 2.

	BEACH	BUS	CAFE	CAR	CITYCENTER	FORESTPATH	GROCERYSTORE	HOME	LIBRARY	METRO	OFFICE	PARK	RESIDENT	TRAIN	TRAM
BEACH	2213	14	16	6	38	20	7	46	20	4	17	35	51	7	39
BUS	50	1806	156	23	18	1	27	9	14	4	0	18	24	81	245
CAFE	20	44	1816	6	152	9	154	32	23	32	1	11	21	65	53
CAR	49	29	5	2106	5	0	1	24	4	8	4	1	4	56	183
CITYCENTER	13	8	78	4	2259	4	18	3	7	13	0	7	51	13	14
FORESTPATH	33	0	9	10	18	2245	2	67	10	8	37	41	20	4	7
GROCERYSTORE	27	29	287	3	61	8	1915	51	27	64	2	0	18	33	12
HOME	77	3	82	3	22	75	29	1982	99	15	75	0	15	15	27
LIBRARY	43	9	55	7	18	22	37	332	1764	40	104	15	49	12	24
METRO	5	11	74	3	55	10	91	32	19	2125	19	2	10	14	37
OFFICE	27	0	1	0	0	18	1	82	31	9	2336	23	16	0	2
PARK	91	28	34	1	37	16	20	67	17	6	17	1938	170	43	31
RESIDENT	60	18	67	6	297	52	22	50	32	12	17	156	1660	35	23
TRAIN	33	102	187	26	42	1	21	18	25	9	0	21	11	1758	240
TRAM	50	249	187	73	22	8	41	23	11	16	0	18	16	184	1632
Total	2791	2350	2968	2271	3012	2473	2386	2798	2103	2365	2629	2292	2144	2320	2569

Fig. 5. Confusion matrix of proposed work for each class

Table 2. Cross fold accuracy

Cross fold	Frame based accuracy
Fold1	76%
Fold2	77%
Fold3	79%
Fold4	78%

The proposed model performance is improved by tuning hyper parameters. In deep learning hyper parameter tuning plays major role that shows better accuracy. L1/L2 regularization, increase of hidden neurons, reduced over fitting of the model. The regularization parameters L1 and L2 are both tuned to 0.00001. This will cause many weights to be small. The deep learning model can be optimized manually with an advanced parameter momentum. Momentum modifies the back propagation and helps in avoiding local minima. Even higher momentum may leads to instability. So this can be controlled with a momentum start, momentum ramp, and momentum stable. These parameters are tuned to 0.2, 0.4, and 1e7 respectively. The number of hidden units in each hidden layer is increased from 50 to 200 and the performance is measured by varying the number of epochs upto 500. The learning rate is tuned to 0.01. Acoustic scene classification of proposed model with hyper tuned parameters provides optimized accuracy and it is shown in Table 3 (Fig. 6).

The classification error rate on Training and Validation data obtained is shown in Fig. 7.

The Fig. 8 shows the training and validation logloss of proposed model.

The cross fold evaluation after hyper tuning of proposed system is shown in Table 4.

Table 3. Comparison of optimized accuracy with baseline system

Acoustic scene	Baseline system	Proposed system	Tuned system
Beach	75.3%	88%	90.5%
Bus	71.8%	74%	88.3%
Cafe	57.7%	75%	79.8%
Car	97.1%	85%	91.1%
City center	90.7%	91%	88.5%
Forest path	79.5%	90%	95.2%
Grocery store	58.7%	76%	83.7%
Home	68.6%	79%	75%
Library	57.1%	70%	78.4%
Metro	91.7%	85%	86.4%
Office	99.7%	92%	89.3%
Park	70.2%	78%	85.3%
Resident	64.1%	67%	71.4%
Train	58.0%	71%	74.3%
Tram	81.7%	67%	64.7%

CROSS VALIDATION METRICS - CONFUSION MATRIX ROW LABELS: ACTUAL CLASS: COLUMN LABELS: PREDICTED CLASS

	BEACH	BUS	CAFE	CAR	CITYCENTER	FORESTPATH	GROCERYSTORE	HOME	LIBRARY	METRO	OFFICE	PARK	RESIDENT	TRAIN	TRAM
BEACH	3402	18	13	34	21	61	5	49	24	3	5	55	21	18	30
BUS	17	3060	145	34	13	1	53	6	20	6	0	35	17	86	225
CAFE	16	91	3018	4	134	2	292	19	38	36	4	10	15	85	18
CAR	29	41	21	3417	1	1	2	6	9	1	3	6	1	25	185
CITYCENTER	32	9	166	0	3337	4	21	0	6	38	0	23	101	28	4
FORESTPATH	15	0	12	6	10	3587	1	45	7	1	15	19	38	5	7
GROCERYSTORE	44	22	276	0	29	10	3116	29	60	82	4	6	17	15	12
HOME	77	14	126	8	30	108	31	2765	302	15	80	48	45	19	15
LIBRARY	38	18	77	3	17	26	40	234	2888	41	134	41	95	16	13
METRO	4	13	141	8	29	29	150	29	31	3227	8	15	20	12	17
OFFICE	21	1	10	2	0	75	1	149	60	13	3363	43	25	1	1
PARK	75	27	62	3	20	66	31	24	8	7	36	3187	155	29	6
RESIDENT	168	26	108	16	299	70	31	16	51	26	16	221	2674	21	2
TRAIN	19	122	337	74	27	13	64	11	29	31	0	70	15	2822	164
TRAM	30	371	108	211	28	9	101	24	48	43	2	60	21	271	2433

Fig. 6. Confusion matrix of proposed work with tuned parameters for each class

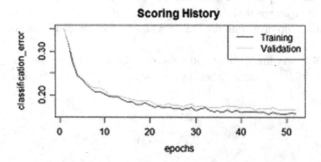

Fig. 7. Classification error rate

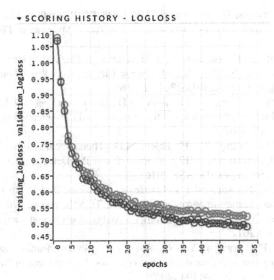

Fig. 8. Logloss

Table 4. Cross folds accuracy

Cross fold	Accuracy
Fold1	82%
Fold2	83%
Fold3	82%
Fold4	82%

4 Conclusion

In this work, a deep learning approach is proposed for Acoustic Scene Classification and it is evaluated with TUT dataset (DCASE 2017). DNN with MFCC features works well better than the baseline system. The accuracy is improved from 74.8% to 82% in the proposed system. In future, the system will be refined in such a way that the classification task works better for all environments with a good degree of flexibility.

References

1. Virtanen, T., Ono, N., Bello, J.P., Glotin, H.: Introduction to the special section on sound scene and event analysis. In: Proceedings of IEEE/ACM Transactions on Audio, Speech, and Language Processing, vol. 25, no. 6, June 2017
2. Bregman, A.S.: Auditory Scene Analysis, International Encyclopedia of the Social and Behavioral Sciences. Pergamon (Elsevier), Amsterdam (1990)

3. Stowell, D., Giannoulis, D., Benetos, E., La-grange, M., Plumbley, M.D.: Detection and classification of acoustic scenes and events. IEEE Trans. Multimedia **17**(10), 1733–1746 (2015)
4. Mesaros, A., Heittola, T., Virtanen, T.: TUT database for acoustic scene classification and sound event detection. In: 24th Acoustic Scene Classification Workshop 2016 European Signal Processing Conference (EUSIPCO) (2016)
5. Kong, Q., Sobieraj, I., Wang, W., Plumbley, M.D.: Deep neural network baseline for Dcase challenge.2016. In: IEEE Proceedings of the Detection and Classification of Acoustic Scenes and Events (DCASE 2016)
6. Xu, Y., Huang, Q., Wang, W., Plumbley, M.D.: Hierarchical learning for DNN-based acoustic scene classification. In: IEEE Proceedings of the Detection and Classification of Acoustic Scenes and Events (DCASE 2016)
7. Schroder, J., Moritz, N., Anemuller, J., Goetze, S., Kollmeier, B.: Classifier architectures for acoustic scenes and events: implications for DNNs, TDNNs, and perceptual features from DCASE 2016. In: IEEE/ACM Transactions on Audio, Speech, and Language Processing, vol. 25, no. 6, June 2017
8. Lee, H., Pham, P., Largman, Y., Ng, A.Y.: Unsupervised feature learning for audio classification using convolutional deep belief networks. In: Advances in Neural Information Processing Systems, pp. 1096–1104 (2009)
9. Laffitte, P., Sodoyer, D., Tatkeu, C., Girin, L.: Deep neural networks for automatic detection of screams and shouted speech in subway trains. In: Proceedings of IEEE International Conference on Acoustics, Speech and Signal Processing (ICASSP), Shanghai, China, pp. 6460–6464, March 2016
10. Valenzise, G., Gerosa, L., Tagliasacchi, M., Antonacci, F., Sarti, A.: Scream and gunshot detection and localization for audio surveillance systems. In: IEEE International Conference on Advanced Video and Signal based Surveillance (2007)
11. Heittola, T., Mesaros, A., Eronen, A., Virtanen, T.: Context-dependent sound event detection. EURASIP J. Audio, Speech Music Process. **1**, 1–13 (2013)
12. Cai, R., Lu, L., Hanjalic, A., Zhang, H., Cai, L.-H.: A flexible framework for key audio effects detection and auditory context inference. IEEE Trans. Audio, Speech Lang. Process. **14**(3), 1026–1039 (2006)
13. Xu, M., Xu, C., Duan, L., Jin, J.S., Luo, S.: Audio keywords generation for sports video analysis. ACM Trans. Multimedia Comput. Commun. Appl. **4**(2), 1–23 (2008)
14. Bugalho, M., Portelo, J., Trancoso, I., Pellegrini, T.S., Abad, A.: Detecting audio events for semantic video search. In: Interspeech, pp. 1151–1154 (2009)
15. Schroder, J., Wabnik, S., van Hengel, P.W.J., Gotze, S.: Detection and classification of acoustic events for in-home care. In: Wichert, R., Eberhardt, B. (eds) Ambient Assisted Living, pp. 181–195. Springer, Heidelberg (2011). https://doi.org/10.1007/978-3-642-18167-2_13
16. Rakotomamonjy, A.: Supervised representation learning for audio scene classification. In: IEEE/ACM Transactions on Audio, Speech, and Language Processing, vol. 25, no. 6, June 2017
17. Goodfellow, I., Bengio, Y., Courville, A.: Deep Learning. MIT Press, Cambridge (2016)
18. Deng, L., Yu, D.: Deep learning: methods and applications. Found. Trends R Signal Process. **7**(34), 197–387 (2014)
19. Li, Y., Zhang, X., Jin, H., Li, X., Wang, Q., He, Q., Huang, A.: Using multi stream hierarchical deep neural network to extract deep audio feature for acoustic event detection. In: Multimedia Tools and Applications. Springer, Berlin (2017). https://doi.org/10.1007/s11042-016-4332-z

20. Li, J., Dai, W., Metze, F., Qu, S., Das, S.: A Comparison of Deep Learning Methods for Environmental Sound Detection (2017)
21. Patiyal, R., Rajan, P.: Acoustic scene classification using deep learning. In: IEEE Proceedings of the Detection and Classification of Acoustic Scenes and Events (DCASE 2016)
22. Dahl, G.E., Sainath, T.N., Hinton, G.E.: Improving deep neural networks for LVCSR using rectifier linear units and dropout. In: 2013 IEEE International Conference on Acoustics, Speech and Signal Processing (ICASSP) (2013)
23. Candel, A., Lanford, J., LeDell, E., Parmar, V., Arora, A.: Deep Learning with H2O, by H2O.ai, c. (2015)

Cyber Security

Certain Challenges in Biometrics System Development

Kamlesh Tiwari[1] and Phalguni Gupta[2(✉)]

[1] Birla Institute of Technology and Science, Pilani, India
[2] National Institute of Technical Teachers' Training and Research, Kolkata, India
phalgunigupta@nitttrkol.ac.in

1 Introduction

Security in today's techno-savvy world where everything is getting digital, is more and more relevant, demanding and challenging. Perpetration of technology due to faster connectivity and low cost computing devices has helped to bring the ubiquitous computing a reality. Digitization has affected every aspect of daily life. Systems for online shopping, social networking, entertainment, gaming, financial transaction *etc.* are made available for 24×7. With the availability of personal and sensitive information available in personal electronic devices, it becomes vital to protect them from malicious access. Security lapses may result in identity thefts or information thefts. Identity thefts correspond to the manipulation of identity whereas the information theft is gaining the copy of secured data.

Authentication is the enabler of security. Specifically in a digital world, the interest is towards systems capable of automatic authentication of an identity. Traditionally these tools are based on token or knowledge. Token provides security of *level 1* which is dependent on something in possession, as an access card, passport, driving license, ATM card, key, dongle *etc.* The user is asked to present the token at the time of authentication. One major drawback of this type of systems is that the security of the token itself is fishy. They can easily be misplaced, lost or stolen. A bit higher level of security, called *level 2*, is offered by techniques using knowledge. Authentication depends upon something known only to the user like personal identification number (PIN) or password. The problem here is again the leakage of the secret. The user for the sake of his reference may like to store the piece of information in some secret place, but the attacker could trace it. Or the user can share the secret information with others. It may also be possible that the user forgets the secret. The adversary can guess easy secrets. Another major problem arise when the user uses the same secret at two different authentication places. There also exist risk of collusion that refers to the possibility of reconstructing fingerprint template even having with partial information. However, there is a possibility of using both *level 1* and *level 2* security means simultaneously, but even then they do not overcome their inherent individual shortcomings.

© Springer Nature Singapore Pte Ltd. 2018
G. Ganapathi et al. (Eds.): ICC3 2017, CCIS 844, pp. 113–123, 2018.
https://doi.org/10.1007/978-981-13-0716-4_10

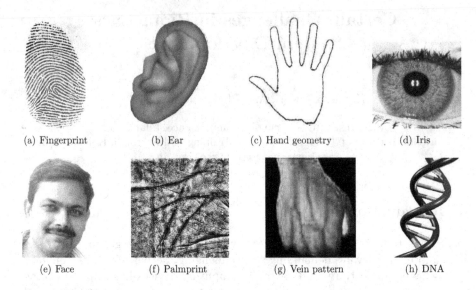

(a) Fingerprint   (b) Ear   (c) Hand geometry   (d) Iris

(e) Face   (f) Palmprint   (g) Vein pattern   (h) DNA

Fig. 1. Some physiological traits

Security measure of *level 3* introduces more complexity to the authentication system by using personal characteristics of the user. It utilizes the physiological or behavioral characteristics, called biometrics, of the individual. It offers a possible solution for identity management with the help of available fully- or semi-automated schemes to recognize an individual. Focus of the work in this thesis is towards biometric based authentication system.

The physiological characteristics like face, fingerprints, iris, ear, palmprint, hand geometry, *etc.* and behavioral characteristics like keystrokes, signature, voice, gait *etc.* are the exclusive properties of a person. These characteristics are referred as traits. It has been found that these properties are not only binding but also are capable of uniquely identifying a person. By using a characteristic feature extraction and its suitable matching technique, one can construct an automatic human recognition system that uses biometric. Some of the advantages of biometric based systems over any token or knowledge based authentication systems are given below.

1. **Unique:** Biometric characteristics are believed to be unique. They are exclusive to an individual and are found to be distinctive enough to identify a person uniquely.
2. **Convenient:** Use of biometric is convenient as the user does not need to carry any authentication token or have to memorize any secret information. Physiological or behavioral characteristics of the user are always available to him which cannot be misplaced, lost or forgotten.
3. **Hard to forge:** Biometric characteristics are hard to forge. An adversary can use a spoofing technique to perform an attack on it, but the simultaneous use of more than one biometric trait massively reduces the chance of forgery.

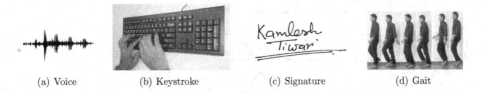

| (a) Voice | (b) Keystroke | (c) Signature | (d) Gait |

Fig. 2. Some behavioral traits

4. **Requires Physical Presence:** A biometric system captures live biometric sample at the time of authentication. Therefore, the physical presence of the user is essential. It inculcates a sense of belongingness with the advantage of non-repudiation; that is, the user cannot deny his participation in authentication at some later point in time.

Some of the physiological and behavioral traits are shown in Figs. 1 and 2 respectively. Each biometric trait has its advantages and disadvantages. The properties of a biometric trait that determine its suitability for the application are

- **Universality:** Every individual must possess the characteristic/attribute for authentication. This property must be present with all and cannot be lost to accident or disease.
- **Permanence:** Biometric characteristics should be constant over an extended period of time. These should not have significant differences in age or disease.
- **Measurability:** The characteristics should be suitable to capture and quantitatively measurable.
- **Acceptability:** The capturing of biometric characteristics should be agreeable to a large class of the population.
- **Uniqueness:** Features in the trait should be distinct from individuals.
- **Performance:** The recognition accuracy, speed and robustness should be within the limitation of the application.
- **Circumvention:** The attribute should be difficult to manipulate or mask. The process should ensure low reproducibility and high reliability.

Biometrics characteristics require various type of imaging technology to capture the evidences. Applications of these broad ranges of biometrics traits vary considerably. In this sub-section, some of the most commonly used physiological traits like face, fingerprint, hand geometry, palmprint, iris, ear, vein pattern, DNA and behavioral traits such as signature, voice, keystroke and gait are discussed.

2   Biometric System Architecture

A biometric system is designed following some pattern recognition techniques. Architecture of a typical biometric system is shown in Fig. 3. It consists of four

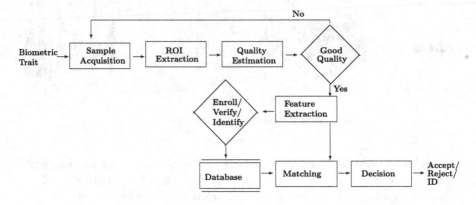

Fig. 3. Sub-modules of a typical biometric system.

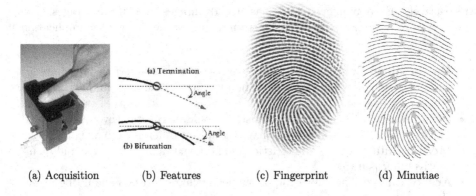

(a) Acquisition   (b) Features   (c) Fingerprint   (d) Minutiae

Fig. 4. Fingerprint acquisition and feature

major tasks that are (i) data acquisition, (ii) preprocessing, (iii) feature extraction, (iv) feature matching and decision (Fig. 4).

- **Data Acquisition:** It is the process of acquiring raw biometrics data using a suitable scanner or sensor. A sensing device such as optical fingerprint scanner, video camera or iris scanner is used for the acquisition. The captured image's quality plays a very important role to generate true features from the trait that are used in matching.
- **Preprocessing:** In this stage, the region of interest (ROI) is extracted from the raw biometric data acquired using filtering, morphological and segmenting operations. Further, the extracted ROI is subjected to enhancement for the correction of non-uniform illumination and to enhance the features.
- **Feature extraction:** In this stage, features which are distinctive information in nature are extracted from the preprocessed trait biometric samples.

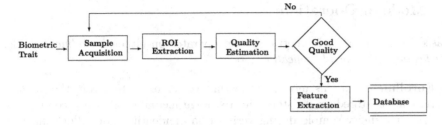

Fig. 5. Enrollment

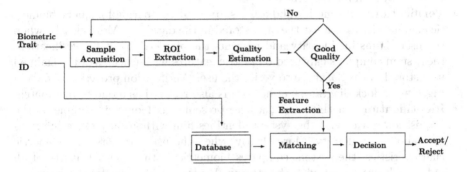

Fig. 6. Verification

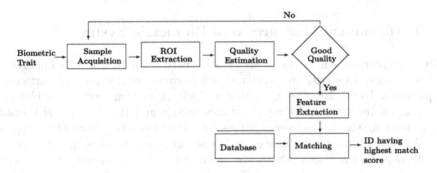

Fig. 7. Identification

These features possess low intraclass and high interclass difference and form a biometrics template.

- **Matching and Decision:** The biometrics template containing features obtained from the test sample are compared against the templates available in the database to make the decision on recognition.

3 Mode of Operation

Based on the need, a biometric system operates on either *Enrollment* mode, *verification* mode or *identification* mode.

- **Enrollment:** In this mode, the feature vector extracted from the biometric trait is stored or registered in a storage medium for future comparison against a query sample during verification or identification. Block diagram of an enrollment process is shown in Fig. 5. This mode registers each subject with the system for recognition.
- **Verification:** This mode validates a subject by comparing his query biometric characteristics against the ones stored in the database. More clearly, when the user claims his identity and presents the biometric trait for verification, the system compares captured biometric features with features of the claimed user stored in the database to verify the user. Verification process is shown in Fig. 6 as a block diagram. Verification is also referred as a one-to-one match.
- **Identification:** In this mode, a user presents his biometric sample at the acquisition device and the system identifies him without any extra information about his identity. More clearly, when the user presents the biometric characteristics, the system compares biometric features with features of all individuals registered with the system. Process of identification is shown in Fig. 7. Identification is also known as one-to-many match.

4 Performance Measures of a Biometric System

Like any other system, performance of a biometric system is a very important factor. It should be critically analyzed before considering it for any particular application. To validate the user using a biometric system, features of biometric samples are matched using a distance metric and the distance is termed as a matching score. However, features extracted from two biometric samples are obtained at different instances and hence may not be the same. It may be due to noise in the sensor, change in biometrics trait due to age, disease, user behavior, change in weather conditions *etc.* Thus, there exists some difference between the two feature vectors. The variability observed in biometric feature of the same individuals is known as *intra-class* variation while the variability in the biometric feature of two different individuals is known as *inter-class* variation. The biometric matching of the same individual is *genuine matching* and the corresponding matching score is *genuine score*. The biometric matching of two different individuals is *imposter matching* and the corresponding score is known as *imposter score*. Matching score can either be a *similarity* or *dissimilarity*. A similarity score represents the correlation between two samples whereas dissimilarity expresses their disagreement. We would be mostly using dissimilarity scores throughout this thesis. If the dissimilarity matching score between two feature vectors is less than an empirically determined threshold, then it is assumed that two corresponding subjects are matched. If a *genuine score* is above the threshold, then the corresponding subject is *falsely rejected*. Similarly if the *imposter score* is below the threshold, then the corresponding subject is *falsely accepted*. Most commonly used measures to evaluate the performance of biometric systems are:

- **False Acceptance Rate (FAR):** For a given threshold, the probability of accepting the imposter as a genuine user is termed as FAR. Increase in the false acceptance rate (FAR) leads to acceptance of unauthorized users that makes the system error prone. A hypothetical FAR curve is shown in Fig. 8.
- **False Rejection Rate (FRR):** For a given threshold, the probability of rejecting the genuine user is termed as FRR. It means, more the value of false reject rate (FRR) more is the falsely rejected genuine subjects. A hypothetical FRR curve is shown in Fig. 9.
- **Equal Error Rate (EER):** It is a point where FAR is equal to FRR and is considered as a measure of verification performance.
- **Failure to Enrol (FTE) Rate:** It is the probability of failure to complete the process of enrolment. It happens when a biometric system is unable to collect and to extract biometric evidences from a person. It may be due to insufficient quality of biometric evidences, improper placement of the bio-metric sample to sensor and sensor problems. Sometimes it is impossible to collect the biometric samples successfully from legitimate users.
- **Receiver Operating Characteristics (ROC) Curve:** This is a graphical plot of FAR against FRR at various thresholds. It is a measure to evaluate the performance of the system for verification mode. A hypothetical ROC curve is shown in Fig. 10.
- **Cumulative Matching Characteristics (CMC) Curve:** Rank-k recognition accuracy R_k is measured as follows:

$$R_k = \frac{N_g}{N_T} \times 100 \tag{1}$$

where N_g denotes the number of genuine matches that occur in top k matches and N_T is the total number of images in the testing set. CMC curve represents the rate of identification accuracy at various ranks R_k.

- **Correct Recognition Rate (CRR):** It is the most commonly used measure to evaluate the performance of the system for identification mode. The CRR is estimated as follows:

$$CRR = \frac{N_1}{N_T} \times 100 \tag{2}$$

where N_1 is the number of genuine matches that occur at top most match or rank-1 and N_T is the total number of images in the testing set. In other words, CRR is recognition accuracy of the top most match.

- **Decidability Index (DI)** signifies how well the separation between genuine and imposter matching scores. If μ_g and μ_i are the mean and σ_g^2 and σ_i^2 are the variances of genuine and imposter distributions respectively then DI is defined by

$$DI = \frac{|\mu_g - \mu_i|}{\sqrt{(\sigma_g^2 - \sigma_i^2)/2}}.$$

Decidability index is found to be high for highly accurate systems.

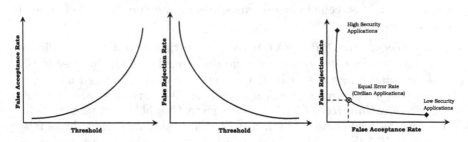

Fig. 8. Hypothetical false acceptance rate curve   **Fig. 9.** Hypothetical false rejection rate curve   **Fig. 10.** Hypothetical ROC curve

- **Hit Rate:** The rate of query images which are retrieved in the candidate list of size t is tyermed as Hit Rate (HR). In other words, if L is the number of queries executed and out of them X number of times the desired item is found to be present in the candidate list, then *Hit Rate* is given by $H_r = (X/L) \times 100\%$.
- **Penetration Rate:** The percentage of database returned as candidate list for a successful retrieval is known as Penetration Rate (PR). If X is the number of query images and d_i is the appropriate number of images retrieved for i^{th} query, then penetration rate is defined by $P_r = \frac{1}{X} \times \sum_{i=1}^{X} d_i \times \frac{1}{N} \times 100\%$ where N is the size of database.

5   Issues in Some Well Known Biometric Traits

Face: Facial image can easily be acquired without being known to the user by using surveillance cameras. This non-invasive data acquisition is one of the main strength of the face over other traits. However, the face recognition systems faces highest number of challenges with respect to the performance. Facial image changes a lot due to change in capture angle, non-cooperative behavior of users, changes in lighting background, illuminations, pose, expressions, occlusions, aging and clutters *etc.* These factors have made the automatic face recognition a very challenging task in computer vision. Face recognition system has two major parts (i) detecting the location of the face in image and (ii) the recognition of face. Holistic face recognition approaches uses the geometric distances between facial characteristics such as eyes, nose, mouth, and forehead as features. Other feature-based approaches use global analysis of the image by representing the face in terms of orthonormal basis vectors.

Fingerprint: It is a well accepted biometric trait for personal authentication because of its reliability, individuality and persistence. A fingerprint is a pattern of ridges and valleys on the surface of finger skin. These patterns are formed during the fetal development and are found to be stable over the lifetime.

Fingerprint patterns differ from person to person even monozygotic twins have different patterns in fingerprints. Fingerprint pattern has been classified into five different classes such as arch, tented arch, left loop, right loop and whorl. Points of discontinuities in ridge are called minutiae and are found to be very useful for recognition. Fingerprint celebrates the advantage of broad acceptance by the public and law enforcement communities. Challenges in the fingerprint based recognition system are posed in terms of getting appropriate number of features (minutiae points), true and only true features and their appropriate quality. The primary disadvantages include needs of high cooperation from the user, prone to spoofing and deterioration in quality of the acquired sample.

Iris: It is considered to be most promising biometric trait yet being most efficient. Iris is the flowery textural pattern in the annular region of the eye that is bounded by the pupil and sclera (the white part of the eye). There exist discriminative textures within iris in the form of furrows, ridges, crypts. Distinctive pattern of complex iris texture makes it suitable for human identification. Automatic iris recognition uses a segmented and fixed size normalized feature vector of iris. Iris faces challenges in terms of segmentation that is extracting only iris part of an eye image. Occlusion due to eyelid and specular reflection also contribute to bad segmentation. The main disadvantage of iris base recognition is the high cost of data acquisition devices and the need for stringent user co-operation for data acquisition.

Hand Geometry: It involves the measurement and analysis of the shape of user's hand. It is a relatively simple procedure, easy to use and inexpensive. Though it requires special hardware to use, it can easily be integrated with other devices or systems. Factors like dry skin do not have any impact on the performance of the system. Further, jewellery or dexterity may pose a challenge in extraction of the correct geometry. Low discriminative capability of hand geometry is one of its major disadvantages.

Palmprint: It i the inner part of a palm image. It contains creases, principal lines, texture, ridges, delta points and local minutiae points *etc.* All these information can be used for recognition. It is found that the palmprint patterns are unique even in monozygotic twins, but the stability of palmprint features is not yet critically studied. The device used to acquire hand image is low cost and user friendly. Area of palmprint is large, therefore it captures more information than a fingerprint. Low cost sensors in a touch-less mannercan be used to capture palmprint image. However, the scanners used to collect the palmprint are bulkier and it needs more processing power due to large capture area. Number of features in an palmprint image is less and due to uneven surface of the palm it is difficult to extract.

Voice: Every individual has distinct voice characteristics like distinct voice texture, pronunciation style etc. Voice recognition attempts to recognize the speaker with the help of pre-stored voice templates. Properties like the fundamental frequency, nasal tone, cadence, inflection, *etc.*, contributes to the identity of the speaker. Voice recognition system consists of three steps (i) recording the acoustic data of a person's voice (ii) converting it to a unique digital signal and (iii) enrolling/matching of the signals. Voice recognition is non-intrusive and has high social acceptability. It also offers a cheap recognition technology because general purpose voice recorders can be used to acquire the data. However, a person's voice can be easily recorded and can be used for unauthorized access. Noise, illness, aging, emotional and physical states of the speaker can change a person's voice, making voice recognition difficult.

Signature: Handwritten signatures are popularly used for offline user authentication in many areas such as banking and legal contracts. Availability of large signature database makes it one of the most needed automatic authentication system. Poor permanence characteristics which is due to the high degree of variability in handwriting with time is one of the major challenges of this technology. An individual's signature can substantially vary over a lifetime. Signature based biometric system records several characteristics from the signature that include the angle of writing, amount of pressure employed, formation of letters, number of connected components, *etc.* However, a signature can easily be reproduced by an imposter. Since everyone may not be able to put signature, lackof universality is also a prime factor. Disadvantages include problems of long-term reliability, lack of performance accuracy and the cost.

Gait: It corresponds to the walking pattern of a user. It is also found to be discriminative user behavior. It can be obtained non-intrusively from distance with a surveillance video camera. Appropriate features such as silhouettes, shape, joint angles, structure and motion are extracted from a video and are used to compare with the stored gait signatures of known individuals. Some systems use the optic flow to describe gait of a person. The disadvantage of gait recognition is that it is computationally expensive and recognition rates are very low compared to other biometrics. Gait is not supposed to be very distinctive across individuals and, therefore it is not well suited for high-security applications. Further, it is not stable and gait is affected due to factors like choice of footwear, clothing and walking surface, *etc.*

Ear: It is being used as a means of human recognition in forensic field for a long time. Like other biometric traits, it contains robust, unique and discriminative features. Shape of the ear is found to be stable and does not change over time. Unlike face, it has no expression. Also, it has uniform color distribution. An ear recognition system is similar to any typical face recognition system. Ear image can also be acquired non-intrusively using cheap sensors. Ear is segmented from the face image. Features obtained from the ear are matched against those stored in a database. Ear has the following advantages: (1) its appearance does not change due to expression, (2) it is found to be unaffected by the aging process, (3) its color is uniform and (4) its background is predictable. The major disadvantage of ear is the occlusion that occurs due to hair or any other foreign body such as earring, cap, earphones *etc.* Ear is very much affected with the pose variations, it is mostly not visible from frontal image. Small size of ear leads to less features.

Knuckleprint: The outer part of a finger is considered as finger knuckle. It contains line based structural features such as knuckle creases, finger wrinkles and texture that are assumed to be discriminative. Use of knuckleprint as a biometric trait is a relatively new approach. Texture and orientation based techniques are popularly used to extract its features and matching. However, a special device may be required to collect the FKP images that may be bulkier and costlier than a fingerprint sensor.

Hand Vein: Vein pattern corresponds to the patterns of blood vessels on the surface of skin. Human vein pattern are found to be invariant over time and are supposed to have discriminant among individuals. An image of vein patterns can be obtained by using an infrared sensor that captures the patterns from the surface of the skin. Vein biometrics has many advantages such as the veins are not visible through the skin, making them extremely difficult to spoof. Shape of the vein is stable it changes very little with as a person grows. Vein recognition has very low error rate and hence, making it suitable for high security applications. The major disadvantage of the hand vein biometrics is that the cost of the infrared sensors and devices used to collect the images is high.

Keystroke: Keystroke dynamics attempts to capture the way people type characters on a keyboard. The style of typing on a computer keyboard is distinguishable from that of another person and is found to be unique enough and that can be used to determine the identity of a person. The distinctive behavioral characteristics measured in keystroke recognition include cumulative typing speed, elapse time between consecutive keystrokes, time of holding each key, the sequence of keys utilized by the person when attempting to type an uppercase or special letter. Typing pattern in keystroke biometrics is generally determined from computer keyboards, But it can also be extracted from any input device such as mobile phones, palmtops, ipad. In comparison to the other biometric technologies, keystroke recognition is the easiest one as it is easy to implement and administer it. There is no need to install any new hardware for it. It can be mentioned here that it is yet to prove the uniqueness of the keystroke recognition and the method has not yet been tested on a large scale.

DNA: Characteristic functionality and the development of living being is controlled by a polymer called deoxyribonucleic acid (DNA). It contains unique biological characteristics of the person that can be used for recognition. Intrusive data acquisition is one of the big disadvantages of DNA. Also, it takes a lot of time to match two DNA samples accurately; therefore this trait is less suitable for online recognition tasks.

6   Conclusions

In this paper, an attempt has been made to raise certain issues to be considered at the time of the development of any biometric system.

A Study on Various Cyber-Attacks and their Classification in UAV Assisted Vehicular Ad-Hoc Networks

N. Vanitha[1,2(✉)] and G. Padmavathi[1]

[1] Department of Computer Science, Avinashilingam Institute for Home Science and Higher Education for Women, University, Coimbatore, Tamil Nadu, India
vanitha969@gmail.com, ganapathi.padmavathi@gmail.com
[2] Department of Information Technology, Dr. N.G.P. Arts and Science College, Coimbatore, Tamil Nadu, India

Abstract. Unmanned Aerial Vehicles (UAV) systems are autonomous systems that can fly separately or it can be functioned remotely without carrying any individuals. These networks prone to various attacks. The people are benefitted from the current growth of networking and cyber world; however, the rapid development of cyber world has furthermore contributed to immoral practices by persons who are using the technology to utilize others. That type of utilization of cyber world with the intension of accessing unauthorized or protected information, collapsing networks, spying, data and currency theft is called as cyber-attack. There is a tremendous increase in Cyber-attacks in number and complexity greater than the previous era, and also lack of awareness on cyber-attacks which has provided many people/societies/groups reveal the true to these attacks. The main aim of this study is to do a brief study of these cyber-attacks in order to create alertness about the various types of attacks and their action so that suitable security methods can be originated against such attacks and also this paper explores the impacts and parameters affected by the most dangerous cyber-attack namely, false data dissemination attack on UAV networks for further implementation and research.

Keywords: Unmanned Aerial Vehicle · Cyberspace · Cyber-attack

1 Introduction

A vehicular ad-hoc network (VANET) provides wireless communication among vehicles and vehicles to road side equipment's. In VANET the frequent path failures, the high mobility, frequently disconnected topology and network traffic density which may affect the reliability of data transmission and routing [2]. These problems are solved using the UAV assisted VANET architecture having U2V/V2U communication. Unmanned Aerial Vehicles (UAV) systems are autonomous systems that can fly separately or it can be functioned remotely without carrying any individuals. UAVs are directly attach to satellites or ground control stations to handover the data. Multiple UAVs can form an ad-hoc network. UAV ad-hoc networks can solve the problems of

G. Ganapathi et al. (Eds.): ICC3 2017, CCIS 844, pp. 124–131, 2018.
https://doi.org/10.1007/978-981-13-0716-4_11

infrastructure-based UAV networks [1]. Nowadays most civil and public applications can be conducted more effectively with multi UAV systems. Figure 1 shows the network model with multi UAVs [9].

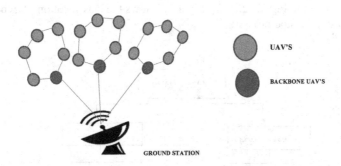

Fig. 1. Network model with multi UAVs

1.1 Applications

UAVs have a great prospective to build numerous applications in both military and civilian domains [1]. Applications include,

- Military applications.
- Civilian applications.

1.2 Research Challenges

There are a multitude of challenges associated with UAV, solving which are crucial for safe and reliable employment of such systems in civilian and military scenarios. The challenges have been identified as being threats to the use of UAV in VANET is given below.

- Reliability
- Traffic density
- Communication among node's
- Conserving energy of a node
- Security. Security plays a major role in UAV to VANET communication.

1.3 Objectives of the Paper

- Presents existing cyber-attacks of the UAV network.
- Presents different cyber-attacks and mode of action with detection techniques.
- Identified Impacts and parameters affected by selected False Data Dissemination attack on UAV networks.
- Security analysis of selected False Data Dissemination Attacks which compromises the security goal on UAV networks are also identified.

2 Major Cyber Attacks

This work particularly focuses on cyber-attacks classifications based on their nature specific to UAV assisted Vehicular ad-hoc networks, here cyber-attacks are categorized into 2 types shown in Fig. 2 [2, 3, 5]. Figure 2 shows the classification of cyber-attacks in UAV assisted ad-hoc network.

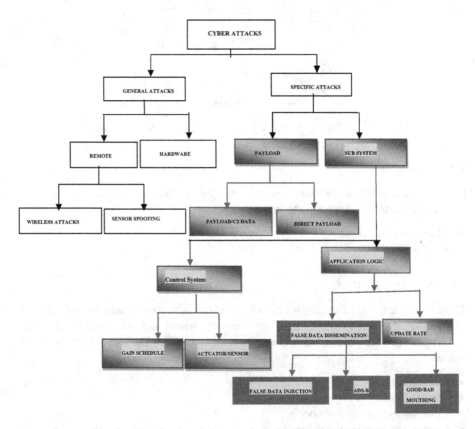

Fig. 2. Classification Cyber-attacks in UAV assisted VANET

2.1 General Attacks

The General attacks are the possible threats and vulnerabilities of the current multi UAV system [4]. General attacks are further classified into two namely [5],

- Hardware attacks
- Remote attack
  - Wireless attacks
  - Sensor spoofing.

The attacker in hardware attacks has access to the hardware components; while, the attackers in wireless attacks use wireless communication path to break through the system, and the attacker in the sensor spoofing, injects false data all the way through the on-board sensors of the UAV autopilot. Wireless or Sensor Jamming/Spoofing attacks comes beneath Remote attack [4].

2.2   Specific Attacks [4]

Specific attacks are the vulnerabilities of the current multi UAV network. Specific attacks are further Classified into two.

Payload Attacks

Payload/C2 Data Attack. This attack "Stealing" Sensor Data, prevalent type of Attack and very easy to do. Gaining Access to the Data stream in order to get "free" intelligence, most streams poorly, or not at all.

Direct Payload Attack. This attack will do Temporary or Permanent damage to Payload, Bit more difficult, but could seriously interrupt operations, Disrupting or destroying the Payload.

Sub System Attacks

These attacks are very difficult to do, but great impact if accomplished [4].

Control System Attack. Prevent H/W or CPU from behaving as programmed Buffer Overflow Exploits, compulsory reset to load malicious code, Hardware Changes or hardware accumulation.

- Gain Schedule Attack. This scheduling attack is used to control non-linear systems through linear systems.
- Actuator/Sensor Attack. This is the way for corrupted data to affect the UAV control by affecting the control surface actuators and associated sensors. This attack can result in failure of control in the UAV.
- *Application Logic Attack.* This attack will do the manipulation of sensors or the environment to provide false data. False data dissemination is one of the application logic attacks. Examples of such attacks are sensor data manipulation, vehicle/component State manipulation, navigation data manipulation, c2 data communication.
- *False data dissemination.* A nasty UAV might program a dissimilar physical noticeable fact like some ecological situations (storm or forest fires) to its neighbors. These attacks can be classified into following sub types,

– False data injection attack
  The attacker introduces fake bandwidth data into routing protocol of UAV network [8]. The main reason of injecting fake bandwidth is to disturb the routing movement of data communication. The fake bandwidth injecting is of two types: the high bandwidth injecting attack and the low bandwidth injecting attack.

Table 1. Different cyber-attacks and mode of action with detection techniques

Name of the Attacks	Description	Mode of action	Techniques
Spoofing attack	GPS handset of a UAV can be spoofed by an hacker	a. GPS Spoofing b. ID, device Spoofing c. Mode confusion	a. RSS streams, summation of detailed coefficients (SDCs) in discrete Hear wavelet transform
False data injection attack	The opponent injects forged bandwidth information into UAV network routing protocol	a. Bandwidth modification b. Payload modification c. The ADS-B attack d. Bad/Good mouthing	a. Adaptive Neural Network b. Hierarchical Identity-Based Signature with Batch Verification
Remote Attack	Attack through one of the sensor or communication channels	a. Wireless Attack b. Sensor Jamming c. Spoofing	a. Adaptive Neural network
Hardware Attack	Access to components directly	a. Out of range b. Crash	a. Monte-Carlo filter
Payload attacks	Stealing Sensor Data	a. Payload/C2 Data Attack b. Direct Payload Attack	a. Time Difference of Arrival (TDOA)
Control System Attack	Attacking the Control System SW or HW	a. Gain Schedule Attack b. Actuator/Sensor Attack c. Buffer overflow d. Load malicious code	a. PI controller b. Delay Adaptive Gains c. Filter innovation d. Adaptive parameter
Application Logic Attack	Altering data to the Control System	a. Sensory & command data manipulation b. State data manipulation c. Navigational and control data manipulation	a. embedded Kalman filter (EKF) b. cyber-attack tolerant (CAT) controller c. Cubature Kalman Filter

- The ADS-B attack (Spoofing GPS Co-ordinates)
  The ADS-B attack [3] tries to disseminate fake data. ADS-B is an on-board element of the UAV structure that televise information's like position and collision evading [3]. From the authors of [3, 4], an ADS-B attacker also tries to broadcast a fake position or takeoff the GPS coordinates (i.e., GPS spoofing) of a board UAV. Because of that, the lifetime of honest drones are exaggerated [3].

Table 2. Impact of False Data Dissemination attack on UAV networks

False Data Dissemination attack	Impact of FDDA attacks	Parameters affected
False Data Injection	• Ground station will get false information about UAV and it disseminate the false information to all UAVs • Communication overhead occurs in the ground station • Packet drop rate increases due to false bandwidth injection • Overall network performance	• Communication Latency • Throughput • Handoff and Roaming • Transmission Robustness • Data traffic, Delay jitter • Bandwidth (data rate)
Spoofing GPS coordinates	• Mobility pattern will be changed • Attacked UAV will be out of range • Not possible to identify the attacked UAV by Ground station or backbone UAV's • Backbone UAV's GPS coordinate get spoofed then whole network information spoofed by others • Particular UAV's survivability affected. • Group survivability affected because of backbone UAV's GPS coordinate spoofing	• Mobility • Throughput • Packet delivery • Handoff and Roaming • Transmission Robustness • Data traffic, Delay jitter • Bandwidth (data rate) • Communication Latency
Good/Bad Mouthing	• Sends false attacker information to UAV's, Backbone UAV's and Ground Station as a response • Not sharing of information • Information not reachable for intended recipient in a group • Information not reachable for an intended group • Information will be taken by malicious node	• Communication Latency • Throughput • Handoff and Roaming • Transmission Robustness • Packet delivery

– Bad/Good mouthing attack
A malicious intrusion detection agent will give the false detection information to corrupt the network performance.

1. Bad-mouthing: UDA declares so as to a good node is malicious
2. Good-mouthing: UDA gives good suggestions concerning a malicious node.

Table 1 summarises the different types of Cyber-attacks in UAV assisted VANET and their mode of action [6–8]. And Table 2 summarises the impacts and parameter affected by False Data Dissemination Attacks on UAV networks.

3 Causes of Attacks Against Security Goals in UAV Network

These attacks cause, increased overhead routing, delays, low packet delivery success rates and traffic. Table 3 summarizes the attacks and its effects on the principles of security on UAV assisted VANET [10, 11].

Table 3. Principles of security and attacks

Attacks/Security Principles	Confidentiality	Integrity	Availability
Spoofing attack	*		*
False data injection attack	*	*	*
Remote Attack	*	*	*
Hardware Attack			*
Payload attacks	*	*	*
Control System Attack	*	*	*
Application Logic Attack	*	*	*

All these attacks could attempt to compromise the availability. Highlighted text shows that Specific subsystem attacks could attempt to compromise all three principles of network security.

4 Conclusion

The handling of computers and Internet engages approximately the entire phases in our life. Cyber security has expanded its impact in modern days. Utilizing cyberspace as glowing, give you an idea about the cyber threats to hack the data of a government administration website; this creates the country covering at the back in their advance deeds. This work focuses on the special types of cyber-attacks and their form of deed with detection techniques in UAV assisted VANET. We have identified security principles that are affected more by specific attacks. Hence, it is required to design an energy efficient intrusion detection system to prevent the security threats and attacks in UAV assisted VANET in future.

References

1. Bekmezci, I., Sahingoz, O.K., Temel, S.: Flying Ad-Hoc Networks (FANETs): a survey. Ad-hoc Netw. **1**, 1254–1270 (2013)
2. Oubbati, O.S., et al.: Intelligent UAV-assisted routing protocol for urban VANETS. Comput. Commun. **107**, 93–111 (2017)
3. Sedjelmaci, H., Senouci, S.M.: A hierarchical detection and response system to enhance security against lethal cyber-attacks in UAV networks. IEEE Trans. Syst. Man Cybern. Syst. **PP**, 1–13 (2016)
4. Kim, A., et al.: Cyber-attack vulnerabilities analysis for unmanned aerial vehicles. In: Infotech@ Aerospace, pp. 1–30 (2012)
5. George, S.: FAA Unmanned Aircraft Systems (UAS) Cyber Security Initiatives, pp. 1–19. Federal Aviation Administration (2015)
6. Podins, K., Stinissen, J., Maybaum, M. (eds.) The vulnerability of UAVs to cyber-attacks - an approach to the risk assessment. In: 5th International Conference on Cyber Conflict (2015)
7. Abbaspoura, A., et al.: Detection of fault data injection attack on UAV using adaptive neural network. Proc. Comput. Sci. **95**, 193–200 (2016)
8. Mokdada, L., Ben-Othmanb, J., Nguyena, A.T.: DJAVAN: detecting jamming attacks in Vehicle Ad hoc Networks. Perform. Eval. **87**, 47–59 (2015)
9. Zhou, Y., Cheng, N., Lu, N., (Sherman) Shen, X.: Multi-UAV-Aided networks. IEEE Veh. Technol. Mag. **10**, 36–44 (2015)
10. He, D., Chan, S., Guizani, M.: Communication security of unmanned aerial vehicles. IEEE Wirel. Commun. **24**, 1415–1420 (2017)
11. Hartmann, K., Steup, C.: The vulnerability of UAVs to cyber-attacks - an approach to the risk assessment. In: Cyber Conflict (CyCon), 5th International Conference, pp. 1–23. IEEE, June 2013

Computational Models

Development of Buoy Information Monitoring System Based on 5G Against the Abandoned, Lost and Discarded Fishing Gears

Kiseon Kim$^{(\boxtimes)}$

MT-IT Collaborations Research Center,
Gwangju Institute of Science and Technology, Gwangju, Korea
kskim@gist.ac.kr

Abstract. Establishment of fishing gear management system becomes an emerging problem, to limit fishing gears (FG) usage and control the abandoned, lost and discarded (ALD) fishing gears. The ALDFG problem confronts investigation of the ownership, type and location information of fishing gears, subsequently, promoting responsible and effective fishing by FG marking, reinforcing the applicability of fisheries control measures including regulations, and effective resources control. This paper introduces a analytic concept and demonstrates an information monitoring system consisting of two parts: one for physical components for the fishery and the other for the ICT-based fishery information monitoring system for efficient fishing gear management. The ICT-based fishery monitoring system includes underwater/surface sensors pertaining to a net, by using Automatic Identification Buoy (AIB), IoT-based network management system, and smart situation-aware visualization.

Keywords: Interdisciplinary computation · 5G communications
Fishery gear · ALD gear · AIB monitoring · Fishery information
FG management

2010 Mathematics Subject Classification: 68U35 (Information Systems)

1 Introduction

In many situations and areas, fishery has been considered for a long time as a free harvesting of available natural foods. Subsequent irresponsible fishing and unsustainable fishery gear deployment confront not only dramatic shortages of ocean food resources but also safety problems due to the unwanted debris and ghost fishing gears [1–4]. Establishment of fishing gear information management system becomes an emerging problem, to restrict fishing gears over-usage and control the abandoned, lost and discarded fishing gears. The UN/FAO declared the Code of Conduct for Responsible Fisheries in 2009, clearly defined the Abandoned, Lost or otherwise Discarded (ALD) Fishing Gear (FG), and further published a report on Combating ALDFG and Ghost Fishing: Development of International Guidelines on the Marking of Fishing Gear in 2016 [5], and there held a series of COFI meetings on Plan for Technical Consultation on ALDFG and IUU.

© Springer Nature Singapore Pte Ltd. 2018
G. Ganapathi et al. (Eds.): ICC3 2017, CCIS 844, pp. 135–143, 2018.
https://doi.org/10.1007/978-981-13-0716-4_12

Accordingly, fishing gear marking standards are expected to be determined around Feb. 2018. Yet, the technical challenges remain to implement the guidelines of the FG marking and subsequent realistic information management [6–9]. As a consequence, development of the FG information management system is expected to enable responsible fishing and sustainable fishery. It's noteworthy that the last decade observed a dramatic evolution in the Information and Computer technologies (ICT), over the broad range from high-speed electronics and source sensor devices, low-cost wideband internet of everything, information visualization based on big data and user experience techniques, to intelligent learning and awareness techniques [10–12]. The sources of the FG information correspond to the terminal sensor devices, typically attached with a buoy, and the information management system monitors the status and information of fishery-centric sensors as variables.

In this paper, we formulate the problem of development of automatic identification buoy (AIB) monitoring system, and propose an analytic approach to 5G-based solutions enabling smart monitoring of underwater/surface monitoring of the Abandoned, Lost and Discarded Fishing Gears.

2 Overall Picture of Fishing Gears Deployment and Fishery Monitoring

Fishing Gear Deployment
Figure 1 depicts one example of cage-type trap nets deployment under fishing, where four underwater acoustic pingers are communicating to the acoustic receivers beneath fishery vessels, and the received acoustic signals are further analyzed to extract the location and status of the fishing gears. Also, the whole net is marked on the surface by using two end floats, namely Buoy. The fishery marking information includes the location and status of each net and gears, which should be eventually conveyed via either acoustic pinger or radio frequency (RF) surface communications, reported to the fishermen or FG owners anywhere and visualized clearly at designated control center. Fishing Gear marking alternatives are shown in the Report on Combating ALDFG and Ghost Fishing [5] in Fig. 2.

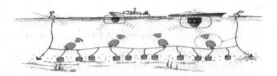

Fig. 1. An example of trap nets under fishery

Some of alternatives in Fig. 2 are valid only for the short distance communications and localization, and buoy type devices (e.g. AIB) are of our interest with long range communications to be deployed with the concept of Internet of Things [11].

Non-technical Issue of FG Development

Properly to convey and deliver the ownership, location and status information regarding the fishery for the visualization and further processing of the monitoring, we should characterize the source data, channel, and available transceiver scheme and devices, and design a monitoring system under realistic environments. Considering that this system design is a typical interdisciplinary field, ICT engineers need close collaborations with fishery scientists and fishermen. Yet, there is not that much statistical information about fishery industries useful for the system implementation. Subsequently, development of FG management starts with the survey of scenarios, analysis and identification of key technical problems as in Fig. 3.

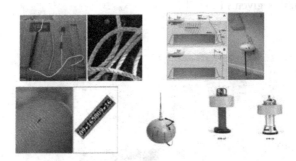

Fig. 2. Alternative sensors and devices for fishing gear marking [5]

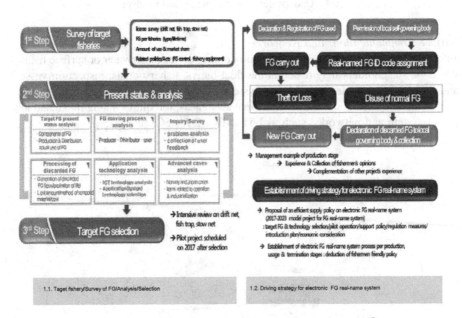

Fig. 3. Master planning of fishing gear management

3   Three Technical Issues of Fishing Gear Management

The fishery monitoring system have multiple objectives, which can be classified into 3 distinctive engineering areas such as (1) underwater localization and communications, (2) surface communication networks, and (3) autonomous situation awareness, and visualization and control.

Underwater Localization and Communications
ALD gears locate at the rocky area around the sea bottom, and the ALDFG should be equipped with acoustic signaling to convey information of the position and status. Especially, low-cost and low-energy designs are very crucial for fishery applications. The pictorial detail is given in Fig. 4.

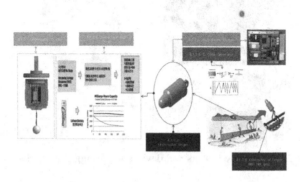

Fig. 4. Underwater acoustic pinters and receivers

Surface Communications and Networks
Information may be gathered from the buoy during either underwater or surface fishing while the stakeholders may reside at the float buoys, fishery boats, inspection vessels and ground fishery/safety monitoring systems. Considering that typical fishery information include fishery status and situational awareness, it requires a long range and low power system and networks, such as LTE NB-IoT for 5G, in Fig. 5.

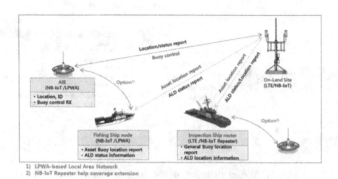

Fig. 5. Basic connections among stakeholders for fishery

Autonomous Situation Awareness and Visualization

Identification, location and status of FG are the basic information for monitoring while fishing or ALD situations, and the information need to be properly conveyed, displayed and controlled for value fishery, especially when unusual events have been acquired, like theft, lost and out-of-control situations. For two scenarios of regular fishing and unusual situations, fishermen's attitudes are quite different for the information monitoring system: the strong against for the former, and the mild for the latter. Considering those former attitudes, visualization and monitoring should seriously consider the privacy issues, not to hurt the property of each fisherman, or, the monitoring system may have trouble while deploying and operating. Further latter, simple improvement of visualization is not sufficient, and more added values should be demonstrated to persuade the meaningful public and safety services of the monitoring system. Prompt notification of any unwanted event, proper localization and tracking of the fishery gear assets, and smart support to handle and carry the unwanted situations, financially and socially.

There exist already several display and control solutions for ocean vessel safety and maritime management system, mainly aware of human lives, vessel safety and container tracing, such as e-Navigation and AIS. Similarly, more things including buoys and fishing gears can be monitored and augmented with prioritized services depending on the design specifications [9]. The specification basically works as it is, and additional low priority things could be served if the communications resources are available, while conventional maritime safety services are guaranteed. Further emergent situations related to fishery could be reactively served on demand.

On the viewpoint of information system operation and management (ONM), the monitoring system consists of two parts: one for physical components for the fishery and the other for the ICT-based fishing gear monitoring system for efficient fishing gear management. Note that the things in Fig. 6 have been conceptually extended to include AIBs, additionally to the AISs.

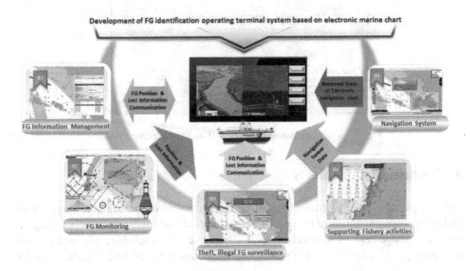

Fig. 6. Visualization stand and an example of display

The physical components are similarly composed of various entities for traditional fishery, such as fishing nets, gears and fishery boats, coastal guard boats and land-based fishery safety management offices. The physical components of fishery already uses VHF communications, Cellular systems, fishery navigation and mapping systems to manage the maritime safety, however, those complex systems are very complicate and inconvenient to use due to its naturally evolution and unorganized patching. Fortunately, e-navigation projects are just imagined to provide seamless and up-to-date maritime services, and the subsequent services will be provided for the fishing gear issues, considering that the abnormal control of the lost and discarded fishing gears leads to maritime safety [15–23].

The ICT-based fishing gear monitoring system includes underwater/surface sensors pertaining to a net, by using Automatic Identification Buoy (AIB), and IoT-based network management system. The acquired information needs to be visualized in such diverse displays as (i) fishermen's smart terminals, (ii) coastal inspectors' pad, (iii) dashboard on land, and (iv) full-scale AIB control center for maritime safety, etc. One example of the display is shown in Fig. 7.

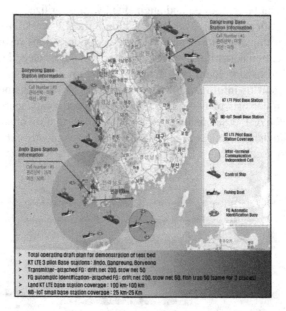

Fig. 7. Verification site plan of AIB and monitoring system

4 Discussion and Effects

Beyond the simple management of existing fishery, based on fishing gear surveillance and theft management technology, it is possible that real-time remote monitoring as well as management of lost/abandoned fishing gear by establishing national control technology. Also, it is possible to carry out the national policy which is the basis for fishing gear management, strengthening policy and fishing gear control law to prevent what is caused by underwater discarded fishing gear, such as fisheries damage, marine ecosystem damage and marine safety accidents [13, 14].

Moreover, economic loss control by monitoring fishing gear management and establishing control system to prevent fishery damage, marine ecosystem damage, and marine safety accidents caused by underwater discarded fishing gear. As part of the national fishing gear management reinforcement policy and fishing gear management law enactment, revitalization of relevant industry and creation of employment by construction of fishing boat, management ship, and land control centers.

In summary, through the monitoring of the fishing gear, the real name system is improved, and through the installation of proper fishing gear, fishing quantity is restricted, so it is possible to prevent over fishing and illegal fishing. Protection of marine ecosystem and reduction of waste collection cost through decreasing, effectively, discarded fishing gears. Reduction of fishing disputes among fishermen. Also, strengthening domestic sovereignty by the management of overlapping fishing gear is within the same fishing ground.

5 Verification and Further Works

Ocean is a new space for Information and computer engineers and it is expected that time and efforts are necessary to make fishermen-friendly monitoring system with allowable ONM cost.

To verify the proposed system, we have to set up real sea experimental scenarios in Fig. 7, for 2 years. The whole optimization of the system will be performed assessing the output of the demonstration, to modify the four key technologies, interactively [24–26].

Further, it is noteworthy that the detailed engineering work to be pursued to implement the low-cost system, as follows:

(1) Development of wireless IoT-based Automatic Identification Buoy.

- low-power·long-distance communication module for Automatic Identification Buoy.
- buoy structure to strengthen waterproofing, protection against weather and sea environments.
- integrated control and embedded hardware and software considering low-power operating.
- built-in, compact and sensitive wireless communication antenna for buoy

(2) Development of lost fishing gear's location information tracking and lost fishing gear management technology

- underwater location transmitter for finding lost fishing gears
- lost fishing gear's location information receiver for fishing vessel.
- automatic identification buoy management technology

(3) Development of situation judgment technology

- fishing gears are of theft and abnormal situation
- AIS buoy integrated with the situation awareness.

6 Conclusive Remarks

This paper formulates the problem of development of fishery information monitoring system with automatic identification buoy (AIB), and proposes an analytic approach to 5G-based solutions enabling smart fishery monitoring of underwater/surface monitoring of the Abandoned, Lost and Discarded Fishing Gears (FG). The FG information management can be supported by 3 distinctive applied mathematics and engineering areas such as (i) underwater localization and communications, (ii) surface communication networks, and (iii) autonomous situation awareness, and information visualization and control. The system is expected to provide information about the ownership, type and location of fishing gears. Subsequently, it will promote responsible and effective fishing with the support of FG marking, reinforce the applicability of fisheries control measures including regulations, and effective resources control for ocean sustainability.

Acknowledgement. This research was a part of the project titled 'Development of Automatic Identification Monitoring System for Fishing Gears', funded by the Ministry of Oceans and Fisheries, Korea.

References

1. Macfadyen, G., Huntington, T., Cappell, R.: Abandoned, lost otherwise discarded fishing gear. FAO Fisheries and Aquaculture Technical Paper No. 523/UNEP Regional Seas Reports and Studies No.185 (2009)
2. Matthews, T.R., Glazer, R.A.: Assessing opinions on abandoned, lost, or discarded fishing gear in the Caribbean. In: Proceedings of the 62nd Gulf and Caribbean Fisheries Institute, Cumana, Venezuela, pp. 12–22, November 2009
3. Kim, S.-G., Lee, W.-I., Moon, Y.: The estimation of derelict fishing gear in the coastal waters of South Korea: trap and gill-net fisheries. Mar. Policy **46**, 119–122 (2014)
4. WWF Final Project Report, Removal of derelict fishing gear, lost or discarded by fishermen in the Baltic Sea, WWF Poland Foundation (2015)
5. Suuronen, P., et al.: Report on Combating ALDFG and Ghost Fishing: Development of International Guidelines on the Marking of Fishing Gear, FAO, GHOST final meeting Venice (2016)
6. Directorate-general for internal policies, Evaluation of various marker buoy techniques for the European Parliament's Committee on Fisheries "The conflict between static gear and mobile gear in inshore fisheries" (2014)
7. Recommendation ITU-R M.1371-5, Technical characteristics for an automatic identification system using time division multiple access in the VHF maritime mobile frequency band (2014)

8. TTA, Technical report TTAR-11.0023, AIS Class-B Protocol Specification (2011)

9. Ellison, B.: SOTDMA Class B AIS, the "new" middle way? http://www.panbo.com/archives/2015/06/sotdma_class_b_ais_the_new_middle_way.html

10. International Standard, IEC 61162-1, Maritime navigation and radiocommunication equipment and systems, Digital interfaces Part 1: Single talker and multiple listeners, International Electrotechnical Commission, Ed. 3 (2007)

11. Huawei: NB-IoT Enabling New Business Opportunities, white paper (2016)

12. Lee, H., et al.: Target acquisition and tracking for ARPA radar development. J. Navig. Port Res. **39**(4), 307–312 (2015)

13. European Commission, The Integrated Fisheries Data Management (IFDM). http://ec.europa.eu/europe2020/index_en.htm

14. Food and Agriculture Organization of the United Nations. www.fao.org

15. Marport. www.marport.com

16. Northwest Atlantic Fisheries Organization. www.nafo.int/

17. Notus Electronics. www.notus.ca

18. Skan-kontroll. www.scancontrol.no

19. Scanmar. www.scanmar.no

20. Simrad. www.simrad.no

21. Sodena. www.sodena.eu, www.ixblue.com

22. Trygg Matt. iuu-vessels.org

23. Vemco. www.vemco.com

24. Korean Fishery Statistics. http://www.kmi.re.kr/

25. Korean Fishery Union. http://www.suhyup.co.kr/

26. Korean Statistics Portal. http://kosis.kr/

27. Agiwal, M., Roy, A., Saxena, N.: Next generation 5G wireless networks: a comprehensive survey. IEEE Commun. Surv. Tutorials **18**(3), 1617–1655 (2016)

28. Gozalvez, J.: New 3GPP standard for IoT [Mobile Radio]. IEEE Veh. Technol. Mag. **11**(1), 14–20 (2016)

29. Zhou, Z., et al.: Energy-efficient resource allocation for D2D communications underlying cloud-RAN-based LTE-A networks. IEEE Internet Things J. **3**(3), 428–438 (2016)

Violation Resolution in Distributed Stream Networks

David Ben-David[1], Guy Sagy[1], Amir Abboud[1], Daniel Keren[2],
and Assaf Schuster[1(✉)]

[1] Technion - Israel Institute of Technology, Haifa, Israel
assaf@cscs.technion.ac.il
[2] Haifa University, Haifa, Israel

Abstract. Distributed stream networks continuously track the global score of the data and alert whenever a given threshold is crossed. The global score is computed by applying a scoring function over the aggregated streams. However, the sheer volume and dynamic nature of the streams impose excessive communication overhead.

Most recent approaches eliminate the need for continuous communication, by using local constraints assigned at the individual streams. These constraints guarantee that as long as no constraint is violated, the threshold is not crossed, and therefore no communication is necessary. Regrettably, local constraint violations become more and more frequent as the network grows and, in the presence of such violations, communication is inevitable.

In this paper, we show that in most cases the violations can be resolved efficiently. Although our solution requires only a reduced subset of the network streams, finding the minimum resolving set is NP-hard. Through analysis of the probability for resolution, we suggest methods to select the resolving set so as to minimize the expected communication overhead and the expected latency of the process. Experimental results with both synthetic and real-life data sets demonstrate that our methods yield considerable improvements over existing approaches.

1 Introduction

Distributed stream networks have become very common in many fields of technology such as sensor networks [49], analysis of financial time series [70], Web applications [41], and more. In these networks, numerous distributed nodes handle highly dynamic, continuous data streams, and their goal is to detect some global property over the distributed data. A fundamental application in distributed stream networks is *Threshold Monitoring*, the goal of which is to constantly alert whenever the value of a predetermined function, evaluated over the network-wide data, crosses a given threshold. A trivial approach for monitoring is to continuously or periodically centralize all the data, thus transforming a distributed problem into a centralized one. This, however, places intolerable burden on the network.

© Springer Nature Singapore Pte Ltd. 2018
G. Ganapathi et al. (Eds.): ICC3 2017, CCIS 844, pp. 144–171, 2018.
https://doi.org/10.1007/978-981-13-0716-4_13

Considerable research efforts were made to reduce network communication overhead in continuous distributed monitoring, as presented in a recent survey by Cormode [14]. Reviewed solutions include data sketching [15] and data sampling [17] algorithms, in which the minimum required amount of data is centralized in order to approximately detect threshold crossing events. Another approach [3,52,60,64] is to assign local constraints at the individual nodes, such that as long as all the constraints are valid, it is guaranteed that the threshold has not been crossed. The latter approach enables exact detection of threshold crossings, while minimizing communication overhead.

The main challenge in the local constraints approach is to efficiently define the constraints so as to minimize the number of violations over time. However, local constraint violations are bound to happen from time to time due to local behaviour (e.g., reading error, energy deficiency or local interrupts). A *local violation* can sometimes indicate that a *global violation*, i.e., threshold crossing, has occurred, but in most cases it suggests nothing more than a local phenomenon. The process which determines the network status in the presence of local violations is referred to as *violation resolution*.

As the size of the network increases, so does the probability that local violations will occur, and thus the frequent need to resolve them efficiently. The resolution process is required to reduce the overall communication cost while meeting a certain latency expectation.

A few violation resolution algorithms have been presented, mainly for star-shaped networks (that consist of a central coordinator). In [3,52,64] the network status is determined by the data and constraints held by the coordinator and the violating nodes. If additional data are required, then the data for the entire network are collected. As the number of local violations increases, this process imposes onerous communication cost. In order to reduce network overhead, the coordinator in [16,37] waits for several violation reports before collecting the entire network data. This process reduces the overall cost but still requires communicating with all the nodes.

Recently, in [59,60], an incremental violation resolution method was presented, which used randomly chosen subsets of non-violating nodes. However, this method is tailored to their algorithms and no bounds were provided for the expected size of these sets.

In this work we address the problem of violation resolution for local constraint monitoring. We present a general approach that attempts to resolve local violations by polling data from a subset of non-violating nodes, referred to as the *resolving set*. Our goal is to reduce communication cost by detecting a minimum-size resolving set, while maintaining a fair latency. To the best of our knowledge, this is the first time the problem of a minimum-size resolving set is studied. We prove that this problem is NP-hard and suggest heuristic approaches for solving it. Assuming homogeneous data setups, we propose a random method (similar to [59]) and present some theoretical bounds using Hoeffding [32] and Bernstein [65] inequalities. Acknowledging the challenge of heterogeneous data setups, we propose an efficient method using algorithms from graph theory. Our methods were extensively tested over both synthetic and real-life data sets, achieving a substantial reduction in communication cost and latency, in comparison to current algorithms.

This paper is organized into eight sections. In Sect. 2 we discuss related work, and notations and terminology are presented in Sect. 3. In Sect. 4 we present an overview of our generic approach. In Sects. 5 and 6 we present our Random Logarithmic algorithm (RLG) and Maximum Matching Tree algorithm (MMT) for homogeneous and heterogeneous setups, respectively. Finally, in Sects. 7 and 8 we present experimental results and conclusions.

2   Related Work

Resolution of local constraint violations is commonly addressed as a subproblem in threshold monitoring algorithms. Threshold monitoring algorithms over star-shaped networks usually proposed [3,52,64][?,?] a two-phase resolution process. First, the coordinator attempts to complete resolution without involving any nodes other than itself and the violating nodes. If it fails, it tries to resolve the violations by communicating, in a single round, with the entire network. A similar approach was suggested for value monitoring algorithms [16,37]. Communication was somewhat reduced as the coordinator would wait for several violation reports before communicating with the entire network. However, in sufficiently large networks, any algorithm that requires communicating with the entire network would incur high communication cost. Furthermore, this approach exposes a trade-off between the communication cost and the latency to alert about a global violation.

Recently, [59,60] suggested gradually increasing the size of the resolving set in a number of rounds. At each round the resolving set was increased by a single non-violating node [60] or by an exponentially increasing number of non-violating nodes [59]. While these methods are closest our work, the nodes in both were selected at random, assuming a homogeneous data setup. In addition, no bounds were presented for the expected size of the resolving set and for the expected latency of the resolution process.

Threshold monitoring problems have also been researched for tree-shaped [34] and peer-to-peer [68] networks. In these algorithms, nodes communicate according to a predefined overlay communication tree. Their resolution processes consist of multiple rounds, where at each round the size of the resolving set is increased by the set of adjacent (or ancestor) nodes of the current resolving set. While the communication tree efficiently reduces the size of the resolving set, neither algorithm suggests a way to define this tree. In our work we present a construction of an overlay tree structure, tailored for heterogeneous data setups.

Finally, several threshold monitoring algorithms [33,50,52] assumed violation resolution processes, but they were not presented by the authors.

3   Violation Resolution and Minimum Resolving Set

3.1   Problem Definition

We consider a distributed online environment consisting of n remote monitoring nodes $N_1, N_2, ..., N_n$ and a central coordinator node N_C. Nodes communicate through the coordinator node, and direct communication between monitoring nodes is not allowed. We assume that the nodes are synchronized and each monitoring node observes an individual stream of multidimensional data over discrete time. Let v_i^t be the d-dimensional real vector collected by node N_i at time t. We denote this vector as the *local vector* of node N_i. Each node is assigned a weight $\omega_i \in \mathbb{R}$, and we define the weighted average of all the local vectors at time t as the *global vector* v^t, i.e., $v^t = (\sum_{i=1}^n \omega_i v_i^t)/(\sum_{i=1}^n \omega_i)$. Note that the weighted average operator can be replaced by any other commutative and associative operator (e.g., multiplication). For simplicity, and w.l.o.g., in the rest of the paper we assume that $\omega_i = 1$ for every node N_i, such that $v^t = \frac{1}{n} \sum_{i=1}^n v_i^t$.

Given an arbitrary monitoring function $f : \mathbb{R}^d \rightarrow \mathbb{R}$ and a threshold value $\tau \in \mathbb{R}$, the coordinator node needs to constantly alert whenever $f(v^t)$ exceeds (or drops below) τ. This threshold monitoring query can be reduced to a simple *domain constraint* over the global vector. Let S be the entire set of vectors over which the monitoring function doesn't cross the threshold, i.e., $S = \{v \in \mathbb{R}^d | f(v) \leq \tau\}$. The coordinator node needs to alert whenever $v^t \notin S$. We denote S, the domain over which this constraint is satisfied, as the *global safe zone*.

Assume that every monitoring node N_i is associated with a subset of the data space $S_i \subseteq \mathbb{R}^d$, denoted as the *local safe zone* of node N_i, such that the following condition holds:

$$\left(\bigwedge_{i=1}^n v_i \in S_i \right) \rightarrow \frac{1}{n} \sum_{i=1}^n v_i \in S. \tag{1}$$

It follows that an *overall satisfaction* of the local domain constraints (imposed by the local safe zones) over the local vectors would imply that the domain constraint over the global vector is also satisfied. In other words, as long as all the local vectors reside within their respective local safe zones ($v_i^t \in S_i$, $i = 1 \ldots n$), the global vector is guaranteed to reside within the global safe zone ($v^t \in S$). This case requires no knowledge of the local vectors of the monitoring nodes, and thus eliminates any need for communication.

At any time t a node's local vector can deviate from its local safe zone. We refer to this event as a *local violation*. When local violations occur, the violating nodes report their local vectors to the coordinator, which then determines the network status to see if a global violation has occurred. This decision process is referred to as *violation resolution*.

An example of a monitoring system in a 2-dimensional space is presented in Fig. 1. It depicts a snapshot of the system in two consecutive time steps. At

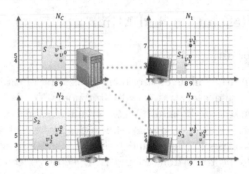

Fig. 1. A monitoring system, consisting of 3 monitoring nodes and a coordinator, in a 2-dimensional space. The safe zones are given as rectangles in the plane and the vectors are marked by dots. It's easy to see that the average of every 3 vectors, taken respectively from the local safe zones, resides within the global safe zone (S). At $t = 0$, all the local vectors reside within their respective safe zones and, consequently, the global vector is also inside the global safe zone. In this case, none of the monitoring nodes reports its local vector to the coordinator. At $t = 1$, a local violation has occurred at N_1 ($v_1^1 \notin S_1$). N_1 would now report its local vector to N_C to seek resolution. N_C must poll N_2 or N_3 (or both) for their local vectors in order to verify that $v^1 \in S$.

the first, all the local constraints are satisfied and the coordinator can determine, without communication, that the global constraint is also satisfied. At the second, a local violation has occurred (at N_1) and resolution is required.

The goal of this work is to reduce the communication required for resolution and altogether maintain fair latency of this process. The latency is the time required to determine the network status, particularly in the case of a global violation. While determining that a global violation has occurred usually requires knowledge of all the local vectors, this case is rare in most monitoring applications (e.g., natural hazards detection). Commonly, if there is no global violation, then the local violations can be resolved by collecting the local vectors of a small set of nodes, referred to as the *resolving set*.

3.2 Resolving Local Violations

We next show how violation resolution is achieved by the resolving set. To this end we generalize the notions of local vector and local safe zone to a set of nodes. Given a set of nodes, A, let v_A^t denote the vector of A at time t and let S_A denote the safe zone of A. v_A^t is defined as the average of the local vectors of the nodes in A at time t, i.e., $v_A^t = \frac{1}{|A|} \sum_{N_i \in A} v_i^t$. Similarly, $S_A = \{\frac{1}{|A|} \sum_{N_i \in A} v_i \mid \bigwedge_{N_i \in A} v_i \in S_i\}$. These definitions are consistent with the operation that defines the global vector (as presented in Sect. 3.1).

Lemma 1. *The above definitions satisfy:*

1. $S_{\{N_1,\dots,N_n\}} \subseteq S$.

2. *For every time t and* mutually disjoint *subsets of nodes $A_1, \ldots, A_m$:*

$$\left(\bigwedge_{i=1}^{m} v_{A_i}^t \in S_{A_i} \right) \rightarrow v_A^t \in S_A$$

where $A = \bigcup_{i=1}^{m} A_i$.

Proof

1. $S_{\{N_1, \ldots, N_n\}} = \{\frac{1}{n} \sum_{i=1}^{n} v_i \mid \bigwedge_{i=1}^{n} v_i \in S_i\}$ by definition. $S_{\{N_1, \ldots, N_n\}} \subseteq S$ follows directly from Eq. 1   □

2. For every time t and mutually disjoint subsets of nodes $A_1, \ldots, A_m$:

$$\bigwedge_{i=1}^{m} v_{A_i}^t \in S_{A_i} \leftrightarrow$$

$$\bigwedge_{i=1}^{m} \left(\frac{1}{|A_i|} \sum_{N_j \in A_i} v_j^t \right) \in \left\{ \frac{1}{|A_i|} \sum_{N_j \in A_i} v_j \mid \bigwedge_{N_j \in A_i} v_j \in S_j \right\} \rightarrow$$

$$\bigwedge_{i=1}^{m} \left(\sum_{N_j \in A_i} v_j^t \right) \in \left\{ \sum_{N_j \in A_i} v_j \mid \bigwedge_{N_j \in A_i} v_j \in S_j \right\} \rightarrow$$

$$\left(\sum_{i=1}^{m} \sum_{N_j \in A_i} v_j^t \right) \in \left\{ \sum_{i=1}^{m} \sum_{N_j \in A_i} v_j \mid \bigwedge_{N_j \in A_i} v_j \in S_j \right\} \rightarrow$$

$$\left(\sum_{N_j \in A} v_j^t \right) \in \left\{ \sum_{N_j \in A} v_j \mid \bigwedge_{N_j \in A} v_j \in S_j \right\} \rightarrow$$

$$\left(\frac{1}{|A|} \sum_{N_j \in A} v_j^t \right) \in \left\{ \frac{1}{|A|} \sum_{N_j \in A} v_j \mid \bigwedge_{N_j \in A} v_j \in S_j \right\} \leftrightarrow$$

$$v_A \in S_A$$

where $A = \bigcup_{i=1}^{m} A_i$.   □

It follows that when local violations occur, the coordinator node can rule out global violation by acquiring only the local vectors of the resolving set. This is justified by the following theorem.

Theorem 1. *Let $\mathcal{V}$ be the entire set of violating nodes at time t. If there exists a set of nodes A such that $\mathcal{V} \subseteq A$ and $v_A^t \in S_A$, then $v^t \in S$.*

Proof. Let $N = \{N_1 \ldots N_n\}$. Since $\mathcal{V} \subseteq A$, $N \backslash A$ consists of merely non-violating nodes, i.e., $v_{\{N_i\}}^t \in S_{\{N_i\}}$ for every $N_i \in N \setminus A$. Since we have also $v_A^t \in S_A$, we conclude by Lemma 1 that $v^t = v_N^t \in S_N \subseteq S$.

Following this theorem, we denote $\mathcal{R} = A \setminus \mathcal{V}$ as the resolving set.

3.3   Running Examples

Next we present two running examples which illustrate the concepts and problems addressed in this work.

EXAMPLE 1 (HOMOGENEOUS DATA). *Assume that air quality sensors are deployed at various geographic locations, measuring the concentration of pollutants the air. Each sensor maintains a vector of its readings, such as the concentrations of NO, and NO_2. We evaluate a function over the vectors of measurements in order to determine the overall air quality, and we wish to detect whenever the air quality drops below a certain threshold. Figure 2a depicts a system consisting of four sensors. The safe zones aim to minimize local violations and are therefore centered around the expected value of the readings in an attempt to cover as much of the distribution area as possible. Normally, the readings of the different nodes, for each pollutant, have the same variance and thus, the safe zones of the different nodes are homogeneous in their dimensions. This implies*

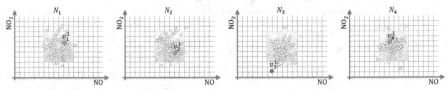

(a) 4-node homogeneous system monitoring concentrations of air pollutants

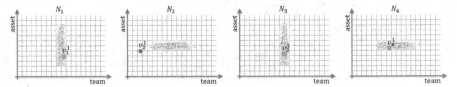

(b) 4-node heterogeneous system monitoring occurences of terms in news reports

Fig. 2. Local violations in homogeneous and heterogeneous systems. The safe zones are given as rectangles in the plane and the local vectors are marked by the enlarged dots. The safe zones were intentionally fit to the distributions of past measurements, represented by the clouds of dots, to minimize local violations. (a) The monitoring nodes are homogeneous in the variance of their distributions and in the dimensions of their safe zones. There is no clear preference for one node over the other in resolving local violations. At $t = 1$, the local violation of N_3 can be resolved by any non-empty subset of the other nodes, and the minimum resolving set comprises a single node. (b) The monitoring nodes are heterogeneous in the variance of their distributions and in the dimensions of their safe zones. Some nodes are more likely to resolve certain local violations than others. At $t = 1$, the local violation of N_2 can only be resolved with N_4, and the minimum resolving set comprises N_4 alone.

that there is no clear preference for one node over the other in resolving local violations.

EXAMPLE 2 (HETEROGENEOUS DATA). *Assume an Internet news agency, which constantly monitors news reports. The nodes are each assigned to a specific category of news (e.g., economy, sports). A wide collection of news-related terms is assembled, and each node tracks the occurrences of the terms in the reports it monitors over a sliding window of one hour. Figure 2b depicts a system consisting of four nodes, of which two are assigned to monitor economic news (N_1, N_3), while the other two monitor sports news (N_2, N_4). The nodes track the occurrences of the terms "team" and "asset." The history of occurrence counts in each node determines its distribution area, and its safe zone is assigned as an interval for each of the terms. As in the air quality example, the safe zones are centered around the expected value and attempt to cover as much of the distribution area as possible. However, in this case, the occurrence counts of each term have a substantially different variance for different nodes; thus, the safe zones of the different nodes are heterogeneous in their dimensions. For example, the term "team" has a wider interval in the sports nodes, while the term "asset" has a wider interval in the economy nodes. This implies, for example, that a sports node has greater flexibility to resolve constraint violations of the term "team."*

4   Generic Algorithm

In this section we present our generic algorithm for violation resolution. In the presence of local violations, the algorithm is executed at the coordinator to determine the network status. The algorithm outputs whether or not the threshold has been crossed. We assume that the coordinator is familiar with the safe zones assigned for the monitoring nodes. Pseudo-code is given in Algorithm 1. At time step t, all violating nodes $(\mathcal{V})$ report their local vectors to the coordinator. Following Theorem 1, the coordinator attempts to resolve the violations by detecting a resolving set $(\mathcal{R})$ such that $v^t_{\mathcal{V} \cup \mathcal{R}} \in S_{\mathcal{V} \cup \mathcal{R}}$. The resolving set is initially empty and gradually extended in every round with the set of nodes returned by the function *getExtendingSet*. The only requirement for this function is to return a non-empty set of non-violating nodes that are not already included in the resolving set. The algorithm terminates when either the violations are resolved (i.e., the threshold was not crossed) or the resolving set comprises the entire set of non-violating nodes. In the latter case, the coordinator directly verifies the network status by evaluating $f(v^t)$. Note that the number of rounds it takes to assemble the resolving set defines the latency of the algorithm, whereas the size of the set defines the communication cost.

Theorem 2. *The generic algorithm always terminates and correctly determines the network status.*

Proof. The while loop (lines 7–16) terminates when either *resolved* = true or $|\mathcal{V} \cup \mathcal{R}| = n$. The properties of getExtendingSet guarantee that the loop will

Table 1. Frequently Used Notations

Notation	Description
n	Number of monitoring nodes
N_i	Monitoring node i ($i = 1 \ldots n$)
N_C	Coordinator node
v_i^t	Vector of node i at time t
v_A^t	Vector of a set of nodes A at time t
v^t	Global vector at time t
S_i	Safe zone of node i
S_A	Safe zone of a set of nodes A
S	Global safe zone
$\mathcal{V}$	Set of violating nodes
$\mathcal{R}$	Resolving set

Algorithm 1 Generic Violation Resolution Algorithm

1: **for all** N_i **in** $\mathcal{V}$ **do**
2:     N_i sends v_i^t to N_C
3: **end for**
4: N_C computes $v_\mathcal{V}^t$, $S_\mathcal{V}$
5: $resolved \leftarrow v_\mathcal{V}^t \in S_\mathcal{V}$ {a boolean flag}
6: $r \leftarrow 1$, $\mathcal{R} \leftarrow \emptyset$
7: **while not** $resolved$ **and** $|\mathcal{V} \cup \mathcal{R}| < n$ **do**
8:     $\mathcal{R}^r \leftarrow$ **getExtendingSet**$(\mathcal{V}, \mathcal{R}, r)$
9:     **for all** N_i **in** $\mathcal{R}^r$ **do**
10:        N_C polls N_i for v_i^t
11:    **end for**
12:    $\mathcal{R} \leftarrow \mathcal{R} \cup \mathcal{R}^r$
13:    N_C computes $v_{\mathcal{V} \cup \mathcal{R}}^t$, $S_{\mathcal{V} \cup \mathcal{R}}$
14:    $resolved \leftarrow (v_{\mathcal{V} \cup \mathcal{R}}^t \in S_{\mathcal{V} \cup \mathcal{R}})$
15:    $r \leftarrow r + 1$
16: **end while**
17: **if not** $resolved$ **then** {$|\mathcal{V} \cup \mathcal{R}| = n$}
18:    $resolved \leftarrow (f(v^t) \leq \tau)$
19: **end if**
20: **return** $resolved$

---

terminate eventually. $resolved$ = true indicates that the violations have been resolved by the resolving set, namely $v_{\mathcal{V} \cup \mathcal{R}}^t \in S_{\mathcal{V} \cup \mathcal{R}}$ (line 5 or 14), and thus, by Theorem 1, a global violation did not occur. Otherwise, $|\mathcal{V} \cup \mathcal{R}| = n$, which suggests that the coordinator holds the local vectors of all the nodes and is therefore able to compute $f(v^t)$ directly (line 18). In any case, by the end of the algorithm, $resolved$ = false if and only if a global violation, i.e., threshold crossing, has occurred. We conclude that the algorithm always terminates and correctly determines the network status.

Throughout the run of the algorithm, the probability for violation resolution, namely $\Pr\{v_{\mathcal{V}\cup\mathcal{R}}^t \in S_{\mathcal{V}\cup\mathcal{R}}\}$, is monotonically non-decreasing. This is implied by the following theorem:

Theorem 3. *Let* $\mathcal{V}, \mathcal{V}^{\mathsf{C}}$ *be the set of violating nodes at time t and its complement* $(\mathcal{V}^{\mathsf{C}} = \{N_1, \ldots, N_n\} \setminus \mathcal{V})$. *Then for every two subsets of nodes* $\mathcal{R}_1 \subseteq \mathcal{R}_2 \subseteq \mathcal{V}^{\mathsf{C}}$:

$$\Pr\left\{v_{\mathcal{V}\cup\mathcal{R}_2}^t \in S_{\mathcal{V}\cup\mathcal{R}_2}\right\} \geq \Pr\left\{v_{\mathcal{V}\cup\mathcal{R}_1}^t \in S_{\mathcal{V}\cup\mathcal{R}_1}\right\}.$$

Proof. Assume that $v_{\mathcal{V}\cup\mathcal{R}_1}^t \in S_{\mathcal{V}\cup\mathcal{R}_1}$ holds. As $\mathcal{V}^{\mathsf{C}}$ is a set consisting of merely non-violating nodes, i.e., $v_i^t \in S_i$ for every $N_i \in \mathcal{V}^{\mathsf{C}}$, it then follows from Lemma 1 that $v_{\mathcal{V}\cup\mathcal{R}_2}^t \in S_{\mathcal{V}\cup\mathcal{R}_2}$. Namely, $v_{\mathcal{V}\cup\mathcal{R}_1}^t \in S_{\mathcal{V}\cup\mathcal{R}_1} \rightarrow v_{\mathcal{V}\cup\mathcal{R}_2}^t \in S_{\mathcal{V}\cup\mathcal{R}_2}$ and hence, $\Pr\{v_{\mathcal{V}\cup\mathcal{R}_2}^t \in S_{\mathcal{V}\cup\mathcal{R}_2}\} \geq \Pr\{v_{\mathcal{V}\cup\mathcal{R}_1}^t \in S_{\mathcal{V}\cup\mathcal{R}_1}\}$.

The performance of the algorithm is dictated exclusively by the function getExtendingSet (line 8), which determines the quality and scale of the extension $\mathcal{R}^r$ to the resolving set. For example, an instance of the generic algorithm, denoted the *Naive Algorithm*, implements this function to always return the set of all non-violating nodes. Consequently, this algorithm achieves the minimum latency (1 round) yet also the maximum communication cost (maximum size resolving set). Another instance, denoted *Random Linear Algorithm* (RLN), extends the resolving set by a single randomly chosen node at each round. While this algorithm attempts to minimize the size of the resolving set, it may incur a rather high number of rounds. These examples expose the trade-off between the expected latency and the expected communication cost of the generic algorithm.

An optimal instance of the generic algorithm is foremost required to reduce communication cost, but at the same time it must maintain a reasonable latency. The latency determines how long the nodes should keep their local vectors and moreover, it determines how long it takes to detect a global violation. This is essential for real-time monitoring applications such as natural hazard detection.

4.1   Minimum Resolving Set Is NP-Hard

Clearly, the optimal communication cost is attained by detecting a minimum size resolving set. However, we argue that even a relaxed version of this problem, in which all the local vectors are known to the coordinator, is NP-hard. Denote this version the minimum resolving set problem (MRS).

Theorem 4. *Let* $\mathcal{V}$ *be the set of all the violating nodes at time t. Given the set of all the local vectors, namely* $\{v_1^t, \ldots, v_n^t\}$, *the problem of finding a resolving set* $\mathcal{R}$ *of a minimum size, such that* $v_{\mathcal{V}\cup\mathcal{R}}^t \in S_{\mathcal{V}\cup\mathcal{R}}$, *is NP-hard.*

Proof. We show a reduction to MRS from the maximum clique problem (MC), which is well-known to be NP-hard. Given a graph $G = (V, E)$, the maximum clique problem is to find a maximum complete subgraph of G, i.e., a set of vertices $V' \subseteq V$ of maximal size, that are pairwise adjacent: $\forall u_i, u_j \in V' : \{u_i, u_j\} \in E$. Given an instance of MC consisting of $G = (V, E), V = \{u_1, \ldots, u_{|V|}\}$, we construct an instance of MRS consisting of $|V| + 1$ monitoring nodes where $\mathcal{V} = \{N_{|V|+1}\}$, i.e., $N_{|V|+1}$ is the single violating node. Let $\{P_1, \ldots, P_m\}$ be the set of all non-adjacent pairs of vertices in G. Namely, $\{u_i, u_j\} \notin E$ if and only if $P_k = \{u_i, u_j\}$ for some $1 \leq k \leq m$. We specify the local vectors and the safe zones in the MRS instance as follows:

- for $i = 1, \ldots, |V|$, $v_i^t = (v_i^t[1], \ldots, v_i^t[m]) \in \mathbb{R}^m$ where:
  $$v_i^t[j] = \begin{cases} 1 & u_i \in P_j \\ 0 & \text{otherwise} \end{cases}.$$
- for $i = 1, \ldots, |V|$:
  $S_i = \{v \in \mathbb{R}^m \mid v[j] \geq 0, \forall j = 1, \ldots, m\}$.
- $v_{|V|+1}^t = v_{\mathcal{V}}^t = 0^m = (0, \ldots, 0)$.
- $S_{|V|+1} = S_{\mathcal{V}} = \{v \in \mathbb{R}^m \mid v[j] > 0, \forall j = 1, \ldots, m\}$.

It is evident that the construction above is polynomial and yields a legal instance of MRS, in which $v_i^t \in S_i$ for $i = 1, \ldots, |V|$ and $v_{\mathcal{V}}^t \notin S_{\mathcal{V}}$. We now show that a solution to the MRS instance, i.e., a minimum resolving set $\mathcal{R} \subseteq \{N_1, \ldots, N_{|V|}\}$, defines a solution of the MC instance, i.e., a maximum clique $V' \subseteq V$, by the following two observations:

1. *A clique V' of size k in the MC instance defines a resolving set $\mathcal{R}$ of size $|V|-k$ in the MRS instance.* We prove that: $\mathcal{R} = \{N_i | u_i \notin V'\}$ is a resolving set. Assume to the contrary that $v_{\mathcal{V} \cup \mathcal{R}}^t \notin S_{\mathcal{V} \cup \mathcal{R}}$. Since $S_{\mathcal{V} \cup \mathcal{R}}$ consists of all strictly positive vectors, there exists $1 \leq j \leq m$ such that $v_{\mathcal{V} \cup \mathcal{R}}^t[j] \leq 0$. Hence, for every $N_i \in \mathcal{R}$, $v_i^t[j] = 0$. This suggests that $P_j \cap \mathcal{R} = \emptyset$ and therefore, $P_j \subseteq V'$. We conclude that V' contains a pair of non-adjacent nodes, a contradiction.
2. *A resolving set $\mathcal{R}$ of size k in the MRS instance defines a clique V' of size $|V| - k$ in the MC instance.* We prove that $V' = \{u_i \mid N_i \notin \mathcal{R}\}$ is a clique. Assume to the contrary that V' is not a clique. Then it contains a pair of non-adjacent nodes. Namely, there exist $1 \leq j \leq m$ such that $P_j \subseteq V'$. Hence, $P_j \cap \mathcal{R} = \emptyset$ and for every $N_i \in \mathcal{R} : v_i^t[j] = 0$. We conclude that $v_{\mathcal{V} \cup \mathcal{R}}^t[j] = 0$ and therefore, $v_{\mathcal{V} \cup \mathcal{R}}^t \notin S_{\mathcal{V} \cup \mathcal{R}}$, a contradiction.

Thus, a minimum resolving set $\mathcal{R}$ in the MRS instance defines a maximum clique V' in the MC instance.

4.2   Probabilistic Analysis of the Algorithm

We next present a few probabilistic bounds which we use to evaluate the expected size of the resolving set. We derive a lower bound on the probability that violation resolution is achieved by the resolving set, namely $\Pr\{v_{\mathcal{V} \cup \mathcal{R}}^t \in S_{\mathcal{V} \cup \mathcal{R}}\}$, and show that it is exponentially increasing in the size of the set. Our method of doing so is to define a region inside $S_{\mathcal{V} \cup \mathcal{R}}$ which contains the expected value of $v_{\mathcal{V} \cup \mathcal{R}}^t$, and then bound the probability that $v_{\mathcal{V} \cup \mathcal{R}}^t$ belongs to this region. We derive two lower bounds for the cases where the bounded region inside the safe zone is a box or a sphere, by employing Hoeffding's and Bernstein's inequalities, respectively.

Hoeffding's Lower Bound – Univariate Data. For simplicity, we first consider the case where the data of each node are one-dimensional ($v_i^t \in \mathbb{R}$). Assume that the safe zone of node N_i is given by an interval on the real line $[a_i, b_i]$ and denote its length by Δ_i. Further, assume that the data of each node are bounded within an interval whose length is $\alpha_i \Delta_i$. It follows that for a set of nodes A, S_A is defined as the interval $[\frac{1}{|A|} \sum_{N_i \in A} a_i, \frac{1}{|A|} \sum_{N_i \in A} b_i]$ whose length is $\frac{1}{|A|} \sum_{N_i \in A} \Delta_i$. Denote by δ^-, δ^+ the distances from the expected value of $v_{\mathcal{V} \cup \mathcal{R}}^t$ to the left and right end points of $S_{\mathcal{V} \cup \mathcal{R}}$, respectively. Let $\mathcal{A} = \mathcal{V} \cup \mathcal{R}$, then:

$$\Pr\{v_{\mathcal{A}}^t \in S_{\mathcal{A}}\} \geq$$
$$\Pr\{\mathbb{E}(v_{\mathcal{A}}^t) - v_{\mathcal{A}}^t \leq \delta^- \wedge v_{\mathcal{A}}^t - \mathbb{E}(v_{\mathcal{A}}^t) \leq \delta^+\} =$$
$$1 - \Pr\{\mathbb{E}(v_{\mathcal{A}}^t) - v_{\mathcal{A}}^t > \delta^- \vee v_{\mathcal{A}}^t - \mathbb{E}(v_{\mathcal{A}}^t) > \delta^+\} =$$
$$1 - \Pr\{\mathbb{E}(v_{\mathcal{A}}^t) - v_{\mathcal{A}}^t > \delta^-\} - \Pr\{v_{\mathcal{A}}^t - \mathbb{E}(v_{\mathcal{A}}^t) > \delta^+\} \geq$$
$$1 - \Pr\{-v_{\mathcal{A}}^t - \mathbb{E}(-v_{\mathcal{A}}^t) \geq \delta^-\} - \Pr\{v_{\mathcal{A}}^t - \mathbb{E}(v_{\mathcal{A}}^t) \geq \delta^+\}.$$

Hoeffding provided an upper bound on the probability for the mean of random variables to deviate from its expected value:

Theorem 5 (Hoeffding 1963, Theorem 2 [32]). *Let $X_1, \ldots, X_n$ be independent random variables such that $a_i \leq X_i \leq b_i$ ($i = 1, \ldots, n$). Then for $t > 0$:*

$$\Pr\{\bar{X} - \mathbb{E}(\bar{X}) \geq t\} \leq \exp\left(-\frac{2n^2t^2}{\sum_{i=1}^n (b_i - a_i)^2}\right)$$

where $\bar{X} = \frac{1}{n} \sum_{i=1}^n X_i$.

We employ Hoeffding's inequality to derive the following corollary:

Corollary 1. *Given that the local vectors of the nodes are univariate and independent, a lower bound on the probability for violation resolution is given by:*

$$\Pr\{v_{\mathcal{V} \cup \mathcal{R}}^t \in S_{\mathcal{V} \cup \mathcal{R}}\} \geq 1 - \phi(\mathcal{V}, \mathcal{R}, \delta^-) - \phi(\mathcal{V}, \mathcal{R}, \delta^+)$$

where

$$\phi(\mathcal{V}, \mathcal{R}, \delta) = \exp\left(-\frac{2(|\mathcal{V}| + |\mathcal{R}|)^2 \delta^2}{\sum_{N_i \in \mathcal{V}} (\alpha_i \Delta_i)^2 + \sum_{N_i \in \mathcal{R}} \Delta_i^2}\right).$$

This bound exponentially approaches 1 as the size of $\mathcal{R}$ increases, regardless of the data distributions of the nodes. Figure 3 depicts these probability bounds for identically distributed nodes.

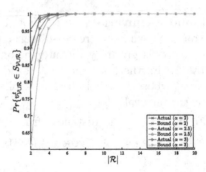

Fig. 3. Hoeffding bounds over univariate synthetic data. In this setup, the safe zone of each node N_i is an interval of length Δ centered around the expected value of v_i^t, and v_i^t is bounded within an interval of length $\alpha\Delta$. Hence, $S_{\mathcal{V}\cup\mathcal{R}}$ is also an interval of length Δ, centered around the expected value of $v_{\mathcal{V}\cup\mathcal{R}}^t$. In addition, $\mathcal{V}$ consists of a single violating node. Therefore, the lower bound for $\Pr\{v_{\mathcal{V}\cup\mathcal{R}}^t \in S_{\mathcal{V}\cup\mathcal{R}}\}$ is given by:
$1 - 2\phi(\mathcal{V}, \mathcal{R}, \Delta/2) = 1 - 2\exp\left(-\frac{(1+|\mathcal{R}|)^2}{2(\alpha^2+|\mathcal{R}|)}\right)$. α denotes how far a node can deviate from its safe zone, in terms of the safe zone's width.

Hoeffding's Lower Bound – Multivariate Data. For the multidimensional case, let $[a_i, b_i] \subseteq \mathbb{R}^d$ be the bounding box of S_i and let $\Delta_i = b_i - a_i$. In other words, the projection of S_i on the jth dimension $(j = 1, \ldots, d)$ is the interval $[a_i[j], b_i[j]]$ of length $\Delta_i[j]$. Assume that the data of each node N_i are bounded within a d-dimensional box: $[c_i, d_i] \subseteq \mathbb{R}^d$, such that $d_i - c_i = \alpha_i \cdot \Delta_i$ where $\alpha_i \in \mathbb{R}^d$ (the product denotes multiplying corresponding entries). Let $\delta^-, \delta^+ \in \mathbb{R}^d$ such that $\delta^-[j], \delta^+[j]$ denote the distances from the expected value of $v_{\mathcal{V}\cup\mathcal{R}}^t$ to the left and right end points of a box bounded in $S_{\mathcal{V}\cup\mathcal{R}}$, respectively, when projected on the jth dimension.

Corollary 2. *Given that the local vectors of the nodes are d-dimensional and independent, a lower bound on the probability for violation resolution is given by:*

$$\Pr\{v_{\mathcal{V}\cup\mathcal{R}}^t \in S_{\mathcal{V}\cup\mathcal{R}}\} \geq$$
$$1 - \sum_{j=1}^{d} \left(\phi(\mathcal{V}, \mathcal{R}, \delta^-, j) + \phi(\mathcal{V}, \mathcal{R}, \delta^+, j)\right)$$

where

$$\phi(\mathcal{V}, \mathcal{R}, \delta, j) = \exp\left(-\frac{2(|\mathcal{V}| + |\mathcal{R}|)^2 \delta[j]^2}{\sum_{N_i \in \mathcal{V}} (\alpha_i[j]\Delta_i[j])^2 + \sum_{N_i \in \mathcal{R}} \Delta_i[j]^2}\right).$$

Bernstein's Lower Bound. An even tighter bound can be attained if we consider a d-sphere of radius δ, inside $S_{\mathcal{V}\cup\mathcal{R}}$ that is centered around the expected value of $v_{\mathcal{V}\cup\mathcal{R}}^t$. Assume that the data of each node are bounded in a d-sphere of radius Δ. Let $\mathcal{A} = \mathcal{V} \cup \mathcal{R}$. Then:

$$\Pr\left\{v_{\mathcal{A}}^t \in S_{\mathcal{A}}\right\} \geq$$
$$\Pr\left\{\|v_{\mathcal{A}}^t - \mathbb{E}(v_{\mathcal{A}}^t)\| \leq \delta\right\} =$$
$$\Pr\left\{\left\|\left(\frac{1}{|\mathcal{A}|}\sum_{N_i \in \mathcal{A}} v_i^t\right) - \mathbb{E}\left(\frac{1}{|\mathcal{A}|}\sum_{N_i \in \mathcal{A}} v_i^t\right)\right\| \leq \delta\right\} =$$
$$\Pr\left\{\left\|\sum_{N_i \in \mathcal{A}} (v_i^t - \mathbb{E}(v_i^t))\right\| \leq \delta|\mathcal{A}|^2\right\} =$$
$$\Pr\left\{\left\|\sum_{N_i \in \mathcal{A}} Z_i\right\| \leq \delta|\mathcal{A}|^2\right\}$$

where $Z_i = v_i^t - \mathbb{E}(v_i^t)$ is a random vector of length d which satisfies $\mathbb{E}(Z_i) = 0$ and $\|Z_i\| < \Delta$. Tropp provided the following generalization of Bernstein's inequalities for a sum of random matrices:

Theorem 6 (Matrix Bernstein [65]). *Given a finite sequence $\{Z_k\}$ of independent, random matrices with dimensions $d_1 \times d_2$, assume that each random matrix satisfies $\mathbb{E}(Z_k) = 0$ and $\|Z_k\| < R$ almost surely. Define:*

$$\sigma^2 := \max\left\{\|\sum_k \mathbb{E}(Z_k Z_k^*)\|, \|\sum_k \mathbb{E}(Z_k^* Z_k)\|\right\}.$$

Then, for all $t \geq 0$,

$$\Pr\left\{\|\sum_k Z_k\| \geq t\right\} \leq (d_1 + d_2) \cdot \exp\left(-\frac{t^2/2}{\sigma^2 + Rt/3}\right).$$

We employ Bernstein's matrix inequality to derive the following corollary:

Corollary 3. *Given that the local vectors of the nodes are d-dimensional and independent, a lower bound on the probability for violation resolution is given by:*

$$\Pr\left\{v_{\mathcal{V}\cup\mathcal{R}}^t \in S_{\mathcal{V}\cup\mathcal{R}}\right\} \geq$$
$$1 - (1+d) \cdot \exp\left(-\frac{(\delta|\mathcal{V}\cup\mathcal{R}|^2)^2/2}{\sigma^2 + \Delta\delta|\mathcal{V}\cup\mathcal{R}|^2/3}\right),$$

where
$$\sigma^2 := \max\left\{\|\sum_{N_i \in \mathcal{V}\cup\mathcal{R}} \mathbb{E}(Z_i Z_i^\mathsf{T})\|, \|\sum_{N_i \in \mathcal{V}\cup\mathcal{R}} \mathbb{E}(Z_i^\mathsf{T} Z_i)\|\right\}.$$

5   Homogeneous Data Instance

In the case of homogeneous data, i.e., the monitoring nodes' data are identically distributed, it appears that no node is clearly preferable over any other in resolving local violations. Therefore, we suggest the following instance of the generic algorithm presented in Sect. 4, to which we refer as the Random Logarithmic algorithm (RLG). Pseudo-code for the function getExtendingSet is given in Algorithm 2. The concept behind this algorithm is straightforward: at each round r the resolving set is extended by additional 2^r randomly selected non-violating nodes.

Following the probabilistic analysis of the algorithm, presented in Sect. 4.2, we are able to estimate the expected size of the resolving set. Thus, we are able to estimate the expected communication cost ($|\mathcal{R}|$), as well as the latency ($\log(|\mathcal{R}|)$) of the algorithm. As depicted in Fig. 3, when the data are homogeneous, the probability for resolution converges rapidly to 1 as the size of the resolving set increases. Therefore, we expect this algorithm to perform well. Note that in the worst case scenario, i.e., the resolving set comprises all the non-violating nodes, RLG bounds the latency by $O(\log(n))$ rounds.

Clearly, the number of nodes added to the resolving set in each round determines the algorithm's latency. We have found that doubling the size of the resolving set at each round yields a fair trade-off between communication cost and latency.

The correctness of RLG derives directly from Theorem 2, as getExtendingSet always returns a non-empty set of non-violating nodes which were not already included in the resolving set.

Algorithm 2 Random Logarithmic Algorithm

---
getExtendingSet($\mathcal{V}, \mathcal{R}, r$)
  1: $\mathcal{R}^r \leftarrow 2^r$ random nodes from $\{N_1, ..., N_n\} \setminus (\mathcal{V} \cup \mathcal{R})$
  2: **return** $\mathcal{R}^r$

---

6   Heterogeneous Data Instance

In various distributed stream networks, the data are heterogeneously distributed, i.e., the distribution of the data may vary greatly among different streams. Data heterogeneity, in most constraint monitoring algorithms, yields heterogeneous constraints. If the constraints are boxes, for example, the safe zones among the different nodes may vary notably in shape, having different width in each dimension. In this section we show that random methods (such as RLG) may produce poor results over heterogeneous setups, and present a more suited algorithm, which efficiently chooses the resolving set. Finally, we discuss the complexity of this algorithm.

6.1   The Heterogeneous Data Challenge

Consider Example 2 presented in Sect. 3.3, in which a single local violation has occurred in one of the sports node (N_2). It is evident that the violation can only be resolved in collaboration with the other sports node (N_4), because economy nodes are characterized by narrow safe zones for sports-related terms. Consequently, they have little or no flexibility in resolving the violation. This is also expressed in the lower bound for the probability of resolution presented in Corollary 2. The flexibility is represented by the width of the safe zone, Δ. It follows that greater flexibility exponentially increases the probability for resolution.

6.2   Maximum Matching Tree Algorithm

We suggest a new instance of the generic algorithm presented in Sect. 4, referred to as the Maximum-Matching Tree algorithm (MMT). The algorithm is preceded by an initialization phase in which an overlay tree structure is defined over the nodes, such that nodes with high probability to resolve each other's local violations are stored under the same sub-tree. The tree is later used in the implementation of getExtendingSet, to retrieve the resolving set.

The construction of the overlay tree, *MMTree*, is presented in detail in the next subsection. The pseudo-code for the construction, as well as the implementation of getExtendingSet, are given in Algorithm 3. Each level in the tree defines a partition of $N_1, \ldots, N_n$, as depicted in Fig. 4. Denote the level of the leaves as 0 and the root level as $\log(n)$. Let $\mathcal{P}^r$ be the partition defined by level r, and let $\mathcal{P}^r[N_i]$ be the set in $\mathcal{P}^r$ that contains the node N_i. In each round r, every node N_i that is not already included in $\mathcal{R}$ and shares the same set in $\mathcal{P}^r$ with a violating node (i.e., $\mathcal{P}^r[N_i] \cap \mathcal{V} \neq \emptyset$) is added to $\mathcal{R}$. Figure 5 illustrates an execution example of MMT over a system of 8 nodes.

The correctness of MMT derives directly from Theorem 2, as getExtendingSet always returns a non-empty set of non-violating nodes which were not already included in the resolving set.

6.3   Maximum Matching Tree Construction

We assume that the coordinator is familiar with the data distributions and the safe zones of all the monitoring nodes. The maximum matching tree is the product of a greedy process that recursively obtains a coarser partition of N by aggregating the components of a finer partition. Given a partition $\mathcal{P}$, of size m, the process obtains a coarser partition $\mathcal{P}'$, of size $\lceil m/2 \rceil$, by *optimally* pairing the components of $\mathcal{P}$. In other words, every component of $\mathcal{P}'$ is formed by joining a pair of components from $\mathcal{P}$ (if m is odd, they will share a single component). Pairing a partition $A_1, \ldots, A_m$ is considered *optimal* if it yields a partition $B_1, \ldots, B_{\lceil m/2 \rceil}$ such that $\Pr\{\bigwedge_{i=1}^{\lceil m/2 \rceil} v_{B_i}^t \in S_{B_i}\}$ is maximized. We initialize the process with the partition of $N = \{N_1, \ldots, N_n\}$ into singletons ($\{N_1\}, \ldots, \{N_n\}$). In turn, optimal pairings are recursively performed until the singleton partition (N) is reached. The result of this is a bottom-up construction of a binary tree where each generated partition defines a new level in the tree. An example of such a tree is depicted in Fig. 4.

Algorithm 3 Maximum-Matching Tree Algorithm

---

buildMMTree()
1: $\mathcal{P}^0 \leftarrow \{\{N_1\}, \ldots, \{N_n\}\}$
2: $r \leftarrow 0$
3: **while** $|\mathcal{P}^r| > 1$ **do**
4:     $\mathcal{G} \leftarrow$ **buildCompleteGraph**$(\mathcal{P}^r)$
5:     $\mathcal{M} \leftarrow$ **findMaximumMatching**$(\mathcal{G})$
6:     $\mathcal{P}^{r+1} \leftarrow \emptyset$
7:     **for all** A in $\mathcal{P}^r$ **do**
8:         Add $\{A \cup \mathcal{M}(A)\}$ to $\mathcal{P}^{r+1}$
9:     **end for**
10:    $r \leftarrow r + 1$
11: **end while**
12: $MMTree \leftarrow \{\mathcal{P}^0, \ldots, \mathcal{P}^r\}$

getExtendingSet$(\mathcal{V}, \mathcal{R}, r)$
1: $\mathcal{R}^r \leftarrow \emptyset$
2: **for all** N_i in $\mathcal{V}$ **do**
3:     Add $\mathcal{P}^r[N_i] \setminus (\mathcal{V} \cup \mathcal{R})$ to $\mathcal{R}^r$
4: **end for**
5: **return** $\mathcal{R}^r$

Optimal Pairing. Given a partition of N into (disjoint) sets, $A_1, \ldots, A_m$, we define a weighted, non-directed, complete graph over these sets. The weight of the edge connecting A_i and A_j is defined by $\log(\Pr\{v^t_{A_i \cup A_j} \in S_{A_i \cup A_j}\})$ for all $1 \leq i < j \leq m$. We perform the pairing by computing a maximum weighted matching in this graph, as described in [18]. As this graph is complete, the matching is perfect (or near-perfect if m is odd). Thus, we obtain a partition, $B_1, \ldots, B_{\lceil m/2 \rceil}$, such that $\sum_{i=1}^{\lceil m/2 \rceil} \log(\Pr\{v^t_{B_i} \in S_{B_i}\})$ is maximized. It follows that $\prod_{i=1}^{\lceil m/2 \rceil} \Pr\{v^t_{B_i} \in S_{B_i}\}$ is maximized and, as the monitoring nodes are independent, we conclude that the pairing is optimal.

Computational Issues. An essential task in the tree construction is computing $\Pr\{v^t_A \in S_A\}$ for a given set of nodes, A. To this end we assume that each node $N_i \in N$ is associated with a probability density function (p.d.f.) given as a discrete set $D_i \subseteq \mathbb{R}^d$ of sample history data. We generalize this notion to a set of nodes A by aggregating the sampled data of the nodes in A (i.e., $\frac{1}{|A|} \sum_{N_i \in A} v_i \in D_A$ where $\{v_i \in D_i\}_{N_i \in A}$ were sampled at the same time steps). It follows that

$$\Pr\{v^t_A \in S_A\} = \frac{|D_A \cap S_A|}{|D_A|}.$$

If the nodes' p.d.f. is given as an explicit function $f_i : \mathbb{R}^d \to [0, 1]$, then $\Pr\{v^t_A \in S_A\} = \int_{S_A} f_A$ where f_A is the convolution of $\{f_i\}_{N_i \in A}$.

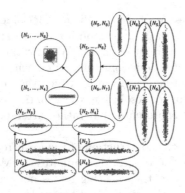

Fig. 4. A maximum matching tree over an 8-node system in a 2-dimensional space. Every level of the tree defines a partition of $\{N_1, \ldots, N_8\}$. The distribution of the average vector and the safe zone are marked by a cloud of dots and a rectangle, respectively, for every partition set. The root node represents the distribution of the global vector and the global safe zone. Note that, indeed, global violations rarely occur. There are two types of nodes: type-1 nodes, which have high variance in the 1st dimension and low variance in the 2nd dimension, and type-2 nodes, which have high variance in the 2nd dimension and low variance in the 1st dimension. 4 nodes of each type comprise the leaves of the tree, denoted by the double outlined ellipses. As expected, MMT first pairs nodes of the same type.

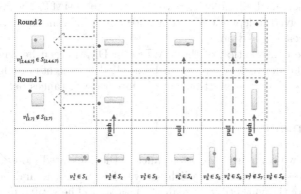

Fig. 5. An execution example of MMT over an 8-node system. At $t = 1$, a snapshot of the system is given at the bottom. First, the violating nodes (N_2, N_7) report their local vectors to the coordinator. Upon failure to resolve the violation, the resolving set is extended with the non-violating nodes from the sets containing N_2, N_7 in the 1st level of the MMTree. If we consider the MMTree in Fig. 4, nodes N_4, N_6 are polled for their local vectors. At this point the violations are resolved and the algorithm terminates.

6.4   Distributed Variant of MMT

Up until now we assumed that the nodes communicate only through the coordinator node. However, this is not necessarily the case in many distributed networks. A star topology in ever expanding networks implies increasing energy costs for the distant nodes. Moreover, as the number of local violations increases with the size of the network, it also implies an increasing load on the coordinator. Next we present a variant of the MMT algorithm, designed for network topologies that support inter-node communication, denoted as Distributed MMT (DMMT). Unlike the generic algorithm, in which the local violations were centralized to the coordinator for resolution, the violating nodes in DMMT do not immediately address the coordinator but rather attempt to resolve their violations locally.

DMMT initiates with the construction of the MMTree by the coordinator as in MMT. In addition, the coordinator disseminates the MMTree to all the nodes. When local violations occur, the violating nodes perform the MMT algorithm simultaneously, yet independently. Each of the nodes constructs its own resolving set using the MMTree until resolution is attained. In each of the sets, the resolution process is led by the smallest index node. It's possible that some of the sets are unified during the process. The MMTree guarantees that, throughout the process, each node can only belong to a single resolving set. In other words, the process creates a partition of the network into disjoint sets of nodes so that according to Lemma 1, the resolution is valid.

The distributed approach dramatically reduces the load on the coordinator. In addition, it can also lead to savings in communication cost. The MMT algorithm attempts to resolve the local violations as a whole and therefore extends the resolving set until all violations are resolved. In DMMT, however, local violations are resolved independently, thus allowing a different size resolving set to be tailored to each violation.

Finally, the construction of the MMTree in DMMT can be adapted to suit the needs of the network. Factors can be applied to weights of edges to reflect desired or non-existing connections and to integrate distances between nodes.

7   Experiments

In this section we compare the performance of the presented violation resolution algorithms. We tested these algorithms over homogeneous and heterogeneous setups, using both synthetic and real-life data sets. The setups used, as well as the performance metrics and compared algorithms, are now described.

7.1   Data Sets

Following are the data sets over which we conducted our experiments:

Syn-HM-n ($n = 16, 32, \ldots, 1024$) – A synthetically generated homogeneous data set consisting of n streams (nodes) of random 3-dimensional data from the normal distribution. The data set was generated such that data variance in each dimension was the same in all the nodes.

Air-HM-n ($n = 16, 32, \ldots, 1024$) – A homogeneous data set taken from the European air quality database (AirBase) [1]. The data set consists of air pollutant measurements read by geographically distributed sensors. The local vectors are

2-dimensional vectors representing the concentrations of NO and NO_2 in the air, which were measured in micrograms per cubic meter. We have assembled n nodes having highly correlated data distributions.

Syn-HT-n ($n = 16, 32, \ldots, 1024$) – A synthetically generated heterogeneous data set consisting of n streams (nodes) of random $\frac{n}{8}$-dimensional data from the normal distribution. The data set was generated such that data variance in each dimension was the same in all the nodes except for 8 nodes in which it was substantially higher.

RCV-HT-n ($n = 16, 32, \ldots, 256$) – The Reuters Corpus (RCV1-v2) [53] consists of 804,014 news stories, each tagged as belonging to one or more of 103 content categories. Every story comprises a precomputed list of terms [48]. We've assembled $\frac{n}{8}$ roughly equally-sized super-categories and selected a term for each category in which it highly dominates the other categories (in the sense that the term occurs many more times in stories within that category). The stories of each super-category were then divided into 8 nodes which tracked the occurrences of all the selected terms over a sliding window of 100 stories (i.e., the local vectors were the occurrence count vectors). This resulted in the nodes having roughly the same variance in every dimension of their data distribution except for the dimension that corresponds to the term of their super-category, in which the variance was substantially higher.

7.2   Scoring Functions and Local Constraints

Following are the scoring functions and local constraints defined for the different data sets:

Syn-HM, Syn-HT, RCV-HT – The global safe zone was defined as a multi-dimensional rectangle (box), centered around the expected value of the global vector. A box essentially sets a lower and an upper bound for the values of the global vector in every dimension. Similarly, the local safe zone of each node was defined as a box centered around the expectation of its data distribution. The boxes were defined such that the average box of the local safe zones was contained in the box of the global safe zone (to ensure that the safe zones condition of Eq. 1 hold). In addition, the boxes achieved a fairly high coverage of the data distribution, guaranteeing that a global violation and a local violation of any node occur with low probability.

Air-HM – The scoring function was defined as the ratio between the average concentrations of NO and NO_2, and the safe zones were defined as triangles – a choice motivated by their simplicity and by their suitability to the data and the definition of the queried function.

Figure 6 provides a graphic illustration of the data sets and safe zones. In the homogeneous data sets the variance of the data in each dimension is approximately the same in all the nodes. Consequently, their safe zones are similar in shape (i.e., have approximately the same width in each dimension). On the other hand, in the heterogeneous data sets the variance of the data in each dimension differs greatly in some nodes and their safe zones are different in shape.

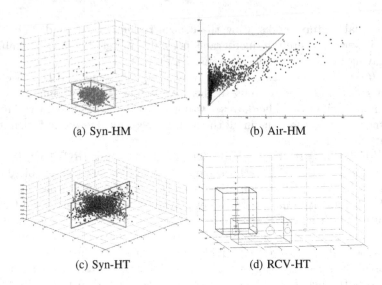

(a) Syn-HM   (b) Air-HM

(c) Syn-HT   (d) RCV-HT

Fig. 6. Data distributions and safe zones of 2 randomly chosen nodes from each data set. In Syn-HT and RCV-HT the data are projected to a 3-dimensional space.

7.3 Performance Metrics

We have applied the following metrics:

Average communication cost – The average number of monitoring nodes that reported their local vector during violation resolution. In the centralized algorithms this corresponds to $\mathcal{V} \cup \mathcal{R}$. In the DMMT algorithm, this excludes the violating nodes which handle the resolution. The actual communication cost (in bytes) is linear in this size.

Average size of resolving set – The average number of non-violating nodes that participated in a violation resolution. This metric emphasizes the overhead in the network resources allocated for resolving the local violations.

Average latency – The average running time (in rounds) of the violation resolution algorithm (namely, the average number of rounds it took to assemble the resolving set).

Average maximum communication load – The maximum communication load is the maximum communication that goes through a single node during violation resolution. In the centralized algorithms this indicates the load on the coordinator.

Note that we've only considered time steps in which a global violation did not occur (i.e., the local violations could be resolved). A global violation would always require any algorithm to collect the entire network data. Moreover, the latency of the algorithms would be maximal (i.e., Naive - 1, RLG/MMT/DMMT - $log(n)$, RLN - n). As global violations are rare, we've omitted them from our evaluation.

7.4   Compared Instances

We evaluate the four instances of the centralized generic algorithm, as well as the distributed version:

Naive – Mentioned in Sect. 4. The resolving set is always defined as the entire set of non-violating nodes.

RLN – Mentioned in Sect. 4. Extends the resolving set linearly with randomly chosen nodes.

RLG – Presented in Sect. 5. Extends the resolving set exponentially with randomly chosen nodes.

MMT – Presented in Sect. 6. Extends the resolving set exponentially using an overlay tree structure.

DMMT – Presented in Subsect. 6.4. A distributed variant of MMT.

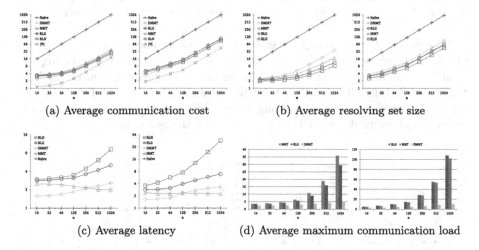

    (a) Average communication cost   (b) Average resolving set size

      (c) Average latency   (d) Average maximum communication load

Fig. 7. Experimental results over Syn-HM (left) and Air-HM (right) homogeneous data sets. The vertical axes of the line graphs are in logarithmic scale. In the average communication cost, all algorithms (except the Naive) approach the minimum, as denoted by the number of violations. The average latency reflects the differences between the algorithms in the expansion rate of the resolving set. DMMT outperforms the centralized algorithms in reducing the average maximum communication load.

7.5   Experimental Results

Homogeneous Setups. We compared the performance of the five algorithms over Syn-HM and Air-HM data sets. Results are presented in Fig. 7. In the average communication cost, RLN, RLG, MMT and DMMT perform similarly and are orders of magnitude away from the Naive algorithm. The graphs of all algorithms, except the naive, closely converge to the graph of the average number of violations. The number of violations defines the minimum communication cost of the resolution, and it increases linearly with the size of the network. This reinforces the hypothesis that in homogeneous setups, no node is clearly preferable to any other, and the choice of the resolving set can be made at random. What really matters is the number of nodes participating in the resolution, as suggested by the lower bounds in Sect. 4.2. The advantage of DMMT, that the violating nodes handle the resolution themselves and don't report their local vectors, is not reflected in the average communication cost. This is explained by the graphs of the average size of the resolving set. Since DMMT resolves each of the local violations independently, it requires more resolving nodes than the centralized algorithms.

In all the algorithms we observe a growth in the graphs of the average size of the resolving set. However, since the number of violating nodes increases, we would expect the number of resolving node to decrease. We reconcile this apparent contradiction by noting that in both data sets, the local violations occur in the positive directions of the axes and thus, they virtually never resolve each other.

The graphs of the average latency reflect the differences between the algorithms in the expansion rate of the resolving set. RLN extends the resolving set linearly and exhibits the worst latency as its graph diverges exponentially. RLN, MMT and DMMT all extend the resolving set exponentially, yet MMT initiates with as many nodes as the violating nodes. The graphs of RLG and DMMT diverge logarithmically while the graph of MMT shows a slow decay.

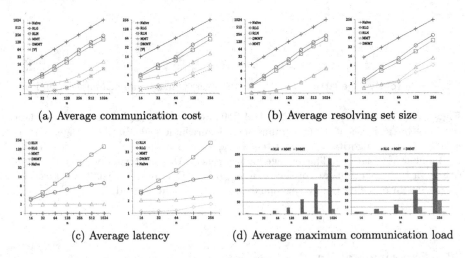

(a) Average communication cost   (b) Average resolving set size

(c) Average latency   (d) Average maximum communication load

Fig. 8. Experimental results over over Syn-HT (left) and RCV-HT (right) heterogeneous data sets. The vertical axes of the line graphs are in logarithmic scale. The clear advantage of MMT and DMMT over the random algorithms is apparent in each of the metrics. DMMT outperforms the centralized algorithms in reducing the average maximum communication load.

As expected, in the average maximum communication load, DMMT outperforms the centralized algorithms, which demonstrate an exponential growth.

Heterogeneous Setups. We compared the performance of the five algorithms over Syn-HT and RCV-HT data sets. Results are presented in Fig. 8. The clear advantage of MMT and DMMT over the random algorithms is apparent in each of the metrics. Due to the diversity of the nodes in the heterogeneous setups, the random algorithms selected nodes that were of no use in the resolution. MMT and DMMT, on the other hand, selected the most relevant nodes according to the preconstructed MMTree. Both MMT and DMMT were successful in reducing the communication cost and the latency almost to the minimum. Nevertheless, DMMT still outperforms MMT in reducing the average maximum communication load. In addition, the savings in reporting the local violations to the coordinator, and the ability to tailor a different size resolving set to each violation independently, may explain the advantage of DMMT in the other metrics as well.

8 Conclusions

This paper focused on minimizing the communication required for handling local constraint violations in distributed threshold monitoring. The key insight was that when there is no global violation, the local violations can typically be resolved without collecting the entire network data.

We presented a formal and precise condition for resolving the violations by a set of nodes. We showed that finding the minimum resolving set is NP-hard, and proposed a general approach that incrementally collects the resolving set. The latency of the process should be taken into account as it determines how long it takes to alert the system about a global violation.

We distinguished between two types of networks: homogeneous and heterogeneous. The network types are related to the correlation between the data distributions of the nodes. We focused especially on the variance of the data in the different dimensions because it can tell us in which direction the node has more tendency to violate. Consequently, it tells us where the safe zone of the node is expected to have greater flexibility in resolving violations. We assumed that in homogeneous networks, no node would be clearly preferable over another in resolving violations. On the other hand, in heterogeneous networks, a careful selection of the resolving nodes can be crucial. These assumptions were reinforced by the lower bounds we presented on the probability for violation resolution. In homogeneous networks, the size of the resolving set was the deciding factor, rather than the identity of its members.

We presented violation resolution algorithms for homogeneous (RLG) and heterogeneous (MMT) setups. Both algorithms guarantee a latency that is logarithmic in the size of the network in the rare case of a global violation. Experimental results with both synthetic and real-life data sets showed that, in homogeneous setups, both algorithms reduced the average communication cost almost to the minimum, and reduced the average latency as well. Due to its simplicity and speed, RLG is preferable. In the heterogeneous setups, however, the superiority of MMT is evident. In addition, if the infrastructure of the network allows it, using DMMT should be considered, in order to avoid the load on the coordinator that is created in the centralized algorithms.

References

1. The European air quality database. http://dataservice.eea.europa.eu/dataservice/
2. Artikis, A., et al.: Scalable proactive event-driven decision making. IEEE Technol. Soc. Magaz. 33, 35–41 (2014)
3. Babcock, B., Olston, C.: Distributed top-k monitoring. In: SIGMOD Conference, New York, NY, USA, pp. 28–39. ACM Press (2003)
4. Bar-Or, A., Keren, D., Schuster, A., Wolff, R.: Hierarchical decision tree induction in distributed genomic databases. IEEE Trans. Knowl. Data Eng. 17(8), 1138–1151 (2005)
5. Bar-Or, A., Schuster, A., Wolff, R., Keren, D.: Decision tree induction in high dimensional, hierarchically distributed databases. In: Proceedings of the 2005 SIAM International Conference on Data Mining, pp. 466–470 (2005)
6. Ben-David, D.: Scalable monitoring of distributed streams. Master Thesis, Technion (2013)
7. Ben-Yehuda, O.A., Ben-Yehuda, M., Schuster, A., Tsafrir, D.: The rise of RaaS: the resource as a service cloud. Commun. ACM 57, 76–84 (2014)
8. Ben-Yehuda, O.A., Posener, E., Ben-Yehuda, M., Schuster, A., Mualem, A.: Ginseng: market-driven memory allocation. ACM SIGPLAN Not. 49, 41–52 (2014)
9. Ben-Yehuda, O.A., Schuster, A., Sharov, A., Silberstein, M., Iosup, A.: Expert: pareto-efficient task replication on grids and a cloud. In: Parallel and Distributed Processing Symposium (IPDPS) (2012)
10. Birk, Y., Keidar, I., Liss, L., Schuster, A.: Efficient dynamic aggregation. In: Dolev, S. (ed.) DISC 2006. LNCS, vol. 4167, pp. 90–104. Springer, Heidelberg (2006). https://doi.org/10.1007/11864219_7
11. Birk, Y., Keidar, I., Liss, L., Schuster, A., Wolff, R.: Veracity radius: capturing the locality of distributed computations. In: Proceedings of the Twenty-Fifth Annual ACM Symposium on Principles of Distributed Computing, pp. 102–111 (2006)
12. Birk, Y., Liss, L., Schuster, A., Wolff, R.: A local algorithm for ad hoc majority voting via charge fusion. In: Guerraoui, R. (ed.) DISC 2004. LNCS, vol. 3274, pp. 275–289. Springer, Heidelberg (2004). https://doi.org/10.1007/978-3-540-30186-8_20
13. Boley, M., Kamp, M., Keren, D., Schuster, A., Sharfman, I.: Communication-efficient distributed online prediction using dynamic model synchronizations. In: BD3@VLDB (2013)
14. Cormode, G.: Continuous distributed monitoring: a short survey. In: Proceedings of the First International Workshop on Algorithms and Models for Distributed Event Processing, AlMoDEP 2011, New York, NY, USA, pp. 1–10. ACM (2011)
15. Cormode, G., Garofalakis, M.N..: Sketching streams through the net: distributed approximate query tracking. In: VLDB Conference, pp. 13–24 (2005)
16. Cormode, G., Muthukrishnan, S., Yi, K.: Algorithms for distributed functional monitoring. ACM Trans. Algorithms 7(2), 21 (2011)

17. Cormode, G., Muthukrishnan, S., Yi, K., Zhang, Q.: Optimal sampling from distributed streams. In: PODS Conference, pp. 77–86 (2010)
18. Edmonds, J.: Paths, trees, and flowers. Canad. J. Math. **17**(3), 449–467 (1965)
19. Friedman, A., Schuster, A.: Data mining with differential privacy. In: Proceedings of the 16th ACM SIGKDD International Conference on Knowledge Discovery and Data Mining, pp. 493–502 (2010)
20. Friedman, A., Schuster, A., Wolff, R.: k-anonymous decision tree induction. In: Fürnkranz, J., Scheffer, T., Spiliopoulou, M. (eds.) PKDD 2006. LNCS (LNAI), vol. 4213, pp. 151–162. Springer, Heidelberg (2006). https://doi.org/10.1007/11871637_18
21. Friedman, A., Sharfman, I., Keren, D., Schuster, A.: Privacy-preserving distributed stream monitoring. In: NDSS (2014)
22. Friedman, A., Wolff, R., Schuster, A.: Providing k-anonymity in data mining. VLDB J. **17**(4), 789–804 (2008)
23. Funaro, L., Ben-Yehuda, O.A., Schuster, A.: Ginseng: market-driven LLC allocation. In: 2016 USENIX Annual Technical Conference (2016)
24. Gabel, M., Keren, D., Schuster, A.: Communication-efficient outlier detection for scale-out systems. In: BD3@VLDB, pp. 19–24 (2013)
25. Gabel, M., Keren, D., Schuster, A.: Monitoring least squares models of distributed streams. In: Proceedings of the 21st ACM SIGKDD International Conference on Knowledge Discovery and Data Mining (2015)
26. Gabel, M., Keren, D., Schuster, A.: Anarchists, unite: practical entropy approximation for distributed streams. In: Proceedings of the 23rd ACM SIGKDD International Conference on Knowledge Discovery and Data Mining, pp. 837–846 (2017)
27. Gabel, M., Schuster, A., Keren, D.: Communication-efficient distributed variance monitoring and outlier detection for multivariate time series. In: Parallel and Distributed Processing Symposium (IPDPS), pp. 37–47 (2014)
28. Giatrakos, N., Deligiannakis, I.S.A., Garofalakis, M., Schuster, A.: Prediction-based geometric monitoring over distributed data streams. In: Proceedings of the 2012 ACM SIGMOD International Conference on Management of Data, pp. 265–276 (2012)
29. Giatrakos, N., Deligiannakis, A., Garofalakis, M., Sharfman, I., Schuster, A.: Distributed geometric query monitoring using prediction models. ACM Trans. Database Syst. (TODS) **39**(2), 16 (2014)
30. Gilburd, B., Schuster, A., Wolff, R.: k-TTP: a new privacy model for large-scale distributed environments. In: Proceedings of the 10th ACM SIGKDD International Conference on Knowledge Discovery and Data Mining, pp. 563–568 (2004)
31. Grumberg, O., Heyman, T., Ifergan, N., Schuster, A.: Achieving speedups in distributed symbolic reachability analysis through asynchronous computation. In: Borrione, D., Paul, W. (eds.) CHARME 2005. LNCS, vol. 3725, pp. 129–145. Springer, Heidelberg (2005). https://doi.org/10.1007/11560548_12
32. Hoeffding, W.: Probability inequalities for sums of bounded random variables. J. Am. Stat. Assoc. **58**, 13–30 (1963)
33. Huang, L., Nguyen, X., Garofalakis, M.N., Jordan, M.I., Joseph, A.D., Taft, N.: In-network PCA and anomaly detection. In: NIPS, pp. 617–624 (2006)
34. Jain, N., Dahlin, M., Zhang, Y., Kit, D., Mahajan, P., Yalagandula, P.: Star: self-tuning aggregation for scalable monitoring. In: VLDB, pp. 962–973 (2007)

35. Kamp, M., Boley, M., Keren, D., Schuster, A., Sharfman, I.: Communication-efficient distributed online prediction by dynamic model synchronization. In: Calders, T., Esposito, F., Hüllermeier, E., Meo, R. (eds.) ECML PKDD 2014. LNCS (LNAI), vol. 8724, pp. 623–639. Springer, Heidelberg (2014). https://doi.org/10.1007/978-3-662-44848-9_40

36. Karmon, K., Liss, L., Schuster, A.: GWiQ-P: an efficient decentralized grid-wide quota enforcement protocol. In: High Performance, Distributed Computing, pp. 222–232 (2005)

37. Keralapura, R., Cormode, G., Ramamirtham, J.: Communication-efficient distributed monitoring of thresholded counts. In: SIGMOD Conference, pp. 289–300 (2006)

38. Keren, D., Sagy, G., Abboud, A., Ben-David, D., Schuster, A., Sharfman, I., Deligiannakis, A.: Geometric monitoring of heterogeneous streams. IEEE Trans. Knowl. Data Eng. **26**, 1890–1903 (2014)

39. Keren, D., Sagy, G., Abboud, A., Ben-David, D., Sharfman, I., Schuster, A.: Safe-zones for monitoring distributed streams. In: BD3@VLDB (2013)

40. Keren, D., Sharfman, I., Schuster, A., Livne, A.: Shape sensitive geometric monitoring. IEEE Trans. Knowl. Data Eng. **24**, 1520–1535 (2012)

41. Kiciman, E., Livshits, V.B.: Ajaxscope: a platform for remotely monitoring the client-side behavior of web 2.0 applications. TWEB **4**(4), 13 (2010)

42. Kolchinsky, I., Schuster, A., Keren, D.: Efficient detection of complex event patterns using lazy chain automata. arXiv preprint arXiv:1612.05110 (2016)

43. Kolchinsky, I., Sharfman, I., Schuster, A.: Lazy evaluation methods for detecting complex events. In: Proceedings of the 9th ACM International Conference on Distributed Event-Based Systems, pp. 34–45 (2015)

44. Krivitski, D., Schuster, A., Wolff, R.: A local facility location algorithm for sensor networks. In: Prasanna, V.K., Iyengar, S.S., Spirakis, P.G., Welsh, M. (eds.) DCOSS 2005. LNCS, vol. 3560, pp. 368–375. Springer, Heidelberg (2005). https://doi.org/10.1007/11502593_28

45. Lazerson, A., Gabel, M., Keren, D., Schuster, A.: One for all and all for one: simultaneous approximation of multiple functions over distributed streams. In: Proceedings of the 11th ACM International Conference on Distributed and Event-Based Systems, pp. 203–214 (2017)

46. Lazerson, A., Keren, D., Schuster, A.: Lightweight monitoring of distributed streams. In: KDD, pp. 1685–1694 (2016)

47. Lazerson, A., Sharfman, I., Keren, D., Schuster, A., Garofalakis, M., Samoladas, V.: Monitoring distributed streams using convex decompositions. In: Proceedings of the VLDB Endowment, vol. 8 (2015)

48. Lewis, D.D., Yang, Y., Rose, T.G., Li, F.: Rcv1: a new benchmark collection for text categorization research. J. Mach. Learn. Res. **5**(4), 361–397 (2004)

49. Madden, S., Franklin, M.J.: Fjording the stream: an architecture for queries over streaming sensor data. In: ICDE, pp. 555–566 (2002)

50. Olston, C., Jiang, J., Widom, J.: Adaptive filters for continuous queries over distributed data streams. In: SIGMOD Conference, pp. 563–574 (2003)

51. Palatin, N., Leizarowitz, A., Schuster, A., Wolff, R.: Mining for misconfigured machines in grid systems. In: Data Mining Techniques in Grid Computing Environments (2008)

52. Rabbat, M., Nowak, R.D.: Distributed optimization in sensor networks. In: IPSN, pp. 20–27 (2004)

53. Rose, T., Stevenson, M., Whitehead, M.: The reuters corpus volume 1 - from yesterday's news to tomorrow's language resources. In: Proceedings of the Third International Conference on Language Resources and Evaluation, Las Palmas de Gran Canaria, May 2002

54. Sagy, G., Keren, D., Sharfman, I., Schuster, A.: Distributed threshold querying of general functions by a difference of monotonic representation. PVLDB **4**(2), 46–57 (2010)

55. Sagy, G., Sharfman, I., Keren, D., Schuster, A.: Top-k vectorial aggregation queries in a distributed environment. J. Parallel Distrib. Comput. **71**(2), 302–315 (2011)

56. Schuster, A., Keren, D., Lazerson, A.: Distributed processing using convex bounding functions. US Patent App. 15/208,721 (2016)

57. Schuster, A., Keren, D., Sagy, G., Sherfman, I.: Method and system of managing and/or monitoring distributed computing based on geometric constraints. Patent number 8949409, application number 12/817,242 (2015)

58. Schuster, A., Wolff, R., Gilburd, B.: Privacy-preserving association rule mining in large-scale distributed systems. In: CCGrid Cluster Computing and the Grid (2004)

59. Sharfman, I., Schuster, A., Keren, D.: Aggregate threshold queries in sensor networks. In: IPDPS, pp. 1–10 (2007)

60. Sharfman, I., Schuster, A., Keren, D.: A geometric approach to monitoring threshold functions over distributed data streams. ACM Trans. Database Syst. **32**(4), 23 (2007)

61. Sharfman, I., Schuster, A., Keren, D.: Aggregate threshold queries in sensor networks. In: Parallel and Distributed Processing Symposium (IPDPS) (2007)

62. Sharfman, I., Schuster, A., Keren, D.: Shape sensitive geometric monitoring. In: Proceedings of the Twenty-Seventh ACM SIGMOD-SIGACT-SIGART Symposium on Principles of Database Systems, pp. 301–310 (2008)

63. Sharfman, I., Schuster, A., Keren, D.: A geometric approach to monitoring threshold functions over distributed data streams. In: Ubiquitous Knowledge Discovery, vol. 6202, pp. 163–186 (2010)

64. Ramamritham, K., Shah, S.: Handling non-linear polynomial queries over dynamic data. In: ICDE Conference (2008)

65. Tropp, J.A.: User-friendly tail bounds for sums of random matrices. ArXiv e-prints, April 2010

66. Verner, U., Schuster, A., Silberstein, M.: Processing data streams with hard real-time constraints on heterogeneous systems. In: Proceedings of the International Conference on Supercomputing, pp. 120–129 (2011)

67. Verner, U., Schuster, A., Silberstein, M., Mendelson, A.: Scheduling processing of real-time data streams on heterogeneous multi-GPU systems. In: Proceedings of the 5th Annual International Systems and Storage Conference (2012)

68. Wolff, R., Bhaduri, K., Kargupta, H.: A generic local algorithm for mining data streams in large distributed systems. IEEE Trans. Knowl. Data Eng. **21**(4), 465–478 (2009)

69. Wolff, R., Schuster, A.: Association rule mining in peer-to-peer systems. In: ICDM Conference, pp. 363–374 (2003)

70. Yi, B.-K., Sidiropoulos, N., Johnson, T., Jagadish, H.V.., Faloutsos, C., Biliris, A.: Online data mining for co-evolving time sequences. In: ICDE, pp. 13–22 (2000)

Fluid Queue Driven by an *M/M/1* Queue Subject to Working Vacation and Impatience

K. V. Vijayashree[1]([✉]) and A. Anjuka[2]

[1] Department of Mathematics, Anna University, Chennai, India
vkviji@annauniv.ac.in
[2] Department of Mathematics, SRM Institute of Science and Technology, Kancheepuram, India

Abstract. In recent years, fluid queueing modelling proves to be very effective in the analysis of computer and communication systems, production inventory systems and many other scenarios. This paper studies a fluid queueing model driven by an *M/M/1* queue subject to working vacation and customer impatience. The fluid in the infinite capacity buffer is assumed to decrease when the background queueing model is empty and increase otherwise. The underlying system of differential difference equations that governs the process are solved using continued fraction and generating function methodologies. Explicit expressions for the joint steady state probabilities of the state of the background queueing model and the content of the fluid buffer are obtained in terms of modified Bessel function of the first kind.

Keywords: Continued fraction · Generating function
Laplace transform · Steady state probabilities
Confluent hypergeometric functions

Mathematics Subject Classification: 60K25 · 90B22

1 Introduction

In many real time situations, the server may become unavailable for a random period of time to perform a secondary task, when there are no customer in the waiting line at the service completion epoch. Such period of server absence is termed as server vacation. Queueing models subject to various vacation policies are of interest to researchers in recent times owing to its wide spread applicability. Queueing models subject to single or multiple exponential vacation are apt to model many practical scenarios [3,7,8]. However, a better modelling assumption would be to assume that the server works at a slower rate during vacation periods in comparison to that of regular working period. Such models are classified as queues subject to working vacations [1,2,12].

© Springer Nature Singapore Pte Ltd. 2018
G. Ganapathi et al. (Eds.): ICC3 2017, CCIS 844, pp. 172–182, 2018.
https://doi.org/10.1007/978-981-13-0716-4_14

Fluid queues has become a fascinating area of research in recent years due to its wide spread applicability in communication systems [4], manufacturing systems [6] etc. Markov modulated fluid queues are a particular class of fluid models useful for modelling many physical phenomenon and they often allow tractable analysis. This paper deals with a particular class of flid queueing model wherein the infinite capacity buffer is modulated by the state of the background queueing model. The modulating process is an *M/M/1* queue subject to working vacation and customer impatience. The source of impatience has always been taken to be wither a long wait already experienced upon arrival at a queue or a long wait anticipated by a customer upon arrival. Server vacation may occur due to several factor like insufficient workload in human behaviour, failure of the server subject to repair period, preventive maintenance period in a production system, secondary tasks assigned to the server (which occurs in computer maintenance and testing), service rendered to arrivals as in priority queueing discipline and so on. For example, consider a production inventory system with impatient customers where a single product is produced at a single facility at a constant rate r_1 per unit time. The demands are assumed to occur at a constant rate d per unit time. When the inventory level is much higher than the demands the facility may shut down some machines to reduce the production speed and hence reduce the holding cost of the inventory. In other words, the facility produce items with a slow speed for a random period of time, say r_2 per unit time. Such a situation also prevails when some of the machines are under maintenance. This is called the working vacation time. Upon arrival, an order is either fulfilled from the inventory if any is available or back-ordered. Customers whose orders are back-ordered become impatient if the demands continue to be met at a lower speed due to the delay in operating all the machines. As a consequence, the orders tend to be cancelled if the customers waiting time exceeds the customer's level of patience. The level of inventory varies from time to time and may be visualised as a fluid process which increases at the rate $r_1 - d$ or $r_2 - d$ if the server is in functional state and decrease at the rate $-d$ when the server remains idle.

The stationary analysis of fluid queueing models in a stochastic environment has been discussed extensively by many authors. For a fluid queue driven by an *M/M/1* queueing model, various techniques have been employed by researchers to obtain the stationary buffer content distribution. Parthasarathy *et al.* [9] presents an explicit expression for the buffer content distribution in terms of modified Bessel function of first kind using continued fraction methodology. Recently, fluid models driven by an *M/M/1* queue subject to various vacation strategies were analysed in steady state by Wang *et al.* [16] and Mao *et al.* [5]. Also, the stationary analysis of a fluid model driven by *M/PH/1* queue and *M/M/c* working vacation queue was analysed by Xu *et al.* [13,14], wherein the probability of empty buffer content and mean of the buffer content were obtained using matrix geometric method. However, in most of the literature relating to fluid queues driven by a vacation queueing models, the buffer content distribution is expressed in the Laplace domain. More recently, Vijayashree and Anjuka [11] presented an explicit expression for the buffer content distribution of a fluid queueing model modulated by an *M/M/1* queue subject to catastrophes and subsequent repair. Also, Ammar [10] obtained the explicit expression for the joint steady state system size probabilities of a fluid queue driven by an *M/M/1* queue with multiple exponential vacation.

The remaining of the paper is organised as follows: Sect. 2 presents the model in detail by introducing the various notations used. Section 3 deals with the detailed derivations of the closed form expressions for the joint state probabilities obtained using continued fraction and generating function methodologies. Section 4 summarizes the work in the form of conclusion. To the best of our knowledge, this paper is the first of its kind to provide an analytical solution for the proposed fluid queueing model.

2   Model Description

Consider a fluid queueing model driven by an *M/M/1* queue subject to working vacation and customer impatience. The customers arrives according to a Poisson process with parameter λ and are serviced according to a exponential distribution with parameter, μ_1. When the system becomes empty, the server begins a vacation of some random duration and the vacation times are assumed to follow exponential distribution with parameter, θ. However, in most of the practical applications, it is appropriate to assume that the server (and hence the system) operates continuously rather than completely becoming idle. Therefore, it is assumed that the customers arriving during the vacation duration are serviced at a slower rate, $\mu_2(< \mu_1)$. Each customer, upon arrival, activates an individual timer, exponentially distributed with parameter, ξ. If the system does not transit to the regular busy period before the customers timer expires, the customer becomes impatient and as a result abandons the queue. Let $N(t)$ denote the number of customers in the system at time t. Define

$$J(t) = \begin{cases} 1, & \text{if the server is busy at time } t. \\ 0, & \text{if the server is on working vacation time } t. \end{cases}$$

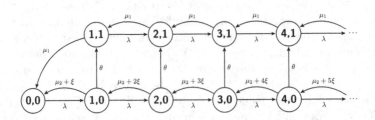

Fig. 1. State Transition Diagram of an *M/M/1* queue subject to working vacation and customer impatience

It is well known that the process $\{(N(t), J(t)), t \geq 0\}$ is a Markov process with the state space

$$\Omega = \{(0,0) \bigcup (k,j), k = 1, 2, \cdots, j = 0, 1\}.$$

The state transition diagram of the background queueing model is given in Fig. 1.

Let $C(t)$ be the content of the buffer at time t. It is assumed that the content of the buffer increases at the rate of r when there are customers in the background queueing model, while the buffer content decreases at the rate r_0 when the system is empty. The dynamics of the buffer content process is given by

$$\frac{dC(t)}{dt} = \begin{cases} 0, & N(t) = 0, C(t) = 0 \\ r_0, & N(t) = 0, C(t) > 0 \\ r, & N(t) > 0, \end{cases}$$

where $r_0 < 0$ and $r > 0$.

Clearly the 3-dimensional process $\{(N(t), J(t), C(t)), t \geq 0\}$ represents a fluid queue driven by an $M/M/1$ queueing model subject to working vacation and customer impatience. The corresponding stability conditions are given by

$$\rho = \frac{\lambda}{\mu_1} < 1 \quad \text{and} \quad d = r_0 \pi_{0,0} + r \sum_{k=1}^{\infty} \pi_{k,0} + r \sum_{k=1}^{\infty} \pi_{k,1} < 0,$$

where $\pi_{k,j}$ represents the steady state probability of the background queueing model to be in state (k, j). The environment process $\{(N(t), J(t)), t \geq 0\}$ is stable if and only if $\rho < 1$. The quantity d is called the mean drift of the process $\{C(t), t \geq 0\}$. When the buffer is infinite, the stochastic process $\{(N(t), J(t), C(t)), t \geq 0\}$ is stable if the mean drift $d < 0$ and $\rho < 1$. When the process $\{(N(t), J(t), C(t)), t \geq 0\}$ is stable, its stationary random vector is denoted by (N, J, C). Under steady state conditions, let

$$F_{k,j}(u) = \lim_{t \to \infty} Pr\{N(t) = k, J(t) = j, C(t) \leq u\}$$
$$= Pr\{N = k, J = j, C \leq u\}, \quad \text{for } u > 0, \text{ and } (k, j) \in \Omega,$$

then the stationary probability distribution of the buffer content C is given by

$$F(u) = P\{C \leq u\} = F_{0,0}(u) + \sum_{k=1}^{\infty} F_{k,0}(u) + \sum_{k=1}^{\infty} F_{k,1}(u).$$

Note that the probability for the content of the buffer to be empty is given by

$$P\{C = 0\} = F_{0,0}(0) + \sum_{k=1}^{\infty} \sum_{j=0}^{1} F_{k,j}(0). \tag{2.1}$$

By standard arguments, the system of differential difference equations that governs the fluid queueing model are given by

$$r_0 \frac{dF_{0,0}(u)}{du} = \mu_1 F_{1,1}(u) - \lambda F_{0,0}(u) + (\mu_2 + \xi)F_{1,0}(u), \tag{2.2}$$

$$r \frac{dF_{k,0}(u)}{du} = \lambda F_{k-1,0}(u) - (\lambda + \mu_2 + n\xi + \theta)F_{k,0}(u) + \mu_2 F_{k+1,0}(u), \ k \geq 1, \tag{2.3}$$

$$r \frac{dF_{1,1}(u)}{du} = \theta F_{1,0}(u) - (\lambda + \mu_1)F_{1,1}(u) + \mu_1 F_{2,1}(u), \tag{2.4}$$

$$r \frac{dF_{k,1}(u)}{du} = \theta F_{k,0}(u) - (\lambda + \mu_1)F_{k,1}(u) + \mu_1 F_{k+1,1}(u) + \lambda F_{k-1,1}(u), \ k \geq 2, \tag{2.5}$$

subject to boundary conditions

$$F_{0,0}(0) = a, \quad F_{k,j}(0) = 0, (k,j) \in \Omega \setminus (0,0) \tag{2.6}$$

The constant a, such that $0 < a < 1$ needs to be determined. The constant a which represents $F_{0,0}(0)$ is determined by adding the Eqs. (2.2) to (2.5) and integrating from zero to infinity. Also, using the boundary conditions represented by Eq. (2.6) in Eq. (2.1) leads to

$$F_{0,0}(0) = a = \frac{(r_0 - r)\pi_{0,0} + r}{r_0}, \tag{2.7}$$

where $\pi_{0,0}$ is given by [15].

3   Stationary Distribution

This section presents the stationary analysis of the proposed fluid queueing model. The stationary distribution of the fluid process play a vital role as they give more information relating to quantities of interest for practical applications like tail probabilities, expected buffer content, traffic intensity, expected delay and sojourn time. The governing system of differential difference equations represented by Eqs. (2.2) to (2.5) are explicitly solved to obtain the joint system state probabilities. The expressions for $F_{k,0}(u), k = 1, 2, \cdots$ are obtain using continued fraction methodology in the Laplace domain and $F_{k,1}(u), k = 0, 1, 2, \cdots$ are solved using generating function methodology.

3.1   Evaluation of $F_{k,0}(u)$

Taking Laplace transform of Eq. (2.3) gives

$$(rs + \lambda + \mu_2 + n\xi + \theta)\hat{F}_{k,0}(s) - [\mu_2 + (n+1)\xi]\hat{F}_{k+1,0}(s) = \lambda \hat{F}_{k-1,0}(s), \quad k = 1, 2, 3, \cdots,$$

which can be written as

$$
\frac{\hat{F}_{k,0}(s)}{\hat{F}_{k-1,0}(s)} = \frac{\lambda}{rs + \lambda + \mu_2 + n\xi + \theta - [\mu_2 + (n+1)\xi]\frac{\hat{F}_{k+1,0}(s)}{\hat{F}_{k,0}(s)}}.
$$

Recursively, it follows that

$$
\frac{\hat{F}_{k,0}(s)}{\hat{F}_{k-1,0}(s)}
$$

$$
= \cfrac{\lambda}{rs + \lambda + \mu_2 + n\xi + \theta - \cfrac{\lambda[\mu_2+(n+1)\xi]}{rs+\lambda+\mu_2+(n+1)\xi+\theta - \cfrac{\lambda[\mu_2+(n+2)\xi]}{rs+\lambda+\mu_2+(n+2)\xi+\theta-[\mu_2+(n+3)\xi]\frac{\hat{F}_{k+3,0}(s)}{\hat{F}_{k+2,0}(s)}}}} \qquad (3.8)
$$

The continued fractions occurring in Eq. (3.8) can be expressed as the ratios of confluent hypergeometric functions. The following is one such identity from Lorentzen and Waadeland [9, (4.1.5), p. 573], which is used in this section:

$$
\frac{{}_1F_1(a+1;c+1;z)}{{}_1F_1(a;c;z)} = \frac{c}{c-z+} \frac{(a+1)z}{c-z+1} \frac{(a+2)z}{c-z+1} \cdots \qquad (3.9)
$$

Using the identity (3.9) in Eq. (3.8) with $a = \frac{\mu_2}{\xi} + k$, $c = \frac{rs+\theta+\mu_v}{\xi} + k$ and $z = \frac{-\lambda}{\xi}$, we get

$$
\frac{\hat{F}_{k,0}(s)}{\hat{F}_{k-1,0}(s)} = \left(\frac{\lambda}{\xi}\right) \frac{{}_1F_1(\frac{\mu_2}{\xi} + k + 1; \frac{rs+\theta+\mu_v}{\xi} + k + 1; \frac{-\lambda}{\xi})}{\left(\frac{rs+\theta+\mu_2}{\xi} + k\right) {}_1F_1(\frac{\mu_2}{\xi} + 1; \frac{rs+\theta+\mu_v}{\xi} + k; \frac{-\lambda}{\xi})}
$$

Repeated application of the above equation for $k = 0, 1, 2, \cdots$ gives

$$
\hat{F}_{k,0}(s) = \hat{\Psi}_k(s)\hat{F}_{0,0}(s) \qquad (3.10)
$$

where

$$
\hat{\Psi}_k(s) = \left(\frac{\lambda}{\xi}\right)^k \frac{1}{\prod_{i=1}^{k}\left(\frac{rs+\theta+\mu_2}{\xi} + i\right)} \frac{{}_1F_1(\frac{\mu_2}{\xi} + k + 1; \frac{rs+\theta+\mu_v}{\xi} + k + 1; \frac{-\lambda}{\xi})}{{}_1F_1(\frac{\mu_2}{\xi} + k; \frac{rs+\theta+\mu_v}{\xi} + k; \frac{-\lambda}{\xi})}
$$

Inversion of Eq. (3.10) yields

$$
F_{k,0}(u) = \Psi_k(u) * F_{0,0}(u) \qquad (3.11)
$$

where $\Psi_k(u)$ and $F_{0,0}(u)$ are explicitly obtained in the subsequent subsections. Thus $F_{k,0}(u)$ for $k = 1, 2, 3, \cdots$ is expressed in terms of $F_{0,0}(u)$.

3.2   Expression for $\Psi_k(u)$

Consider

$$
\hat{\Psi}_k(s) = \left(\frac{\lambda}{\xi}\right)^k \frac{1}{\prod_{i=1}^{k}\left(\frac{rs+\theta+\mu_2}{\xi} + i\right)} \frac{{}_1F_1(\frac{\mu_2}{\xi} + k + 1; \frac{rs+\theta+\mu_v}{\xi} + k + 1; \frac{-\lambda}{\xi})}{{}_1F_1(\frac{\mu_2}{\xi} + 1; \frac{rs+\theta+\mu_v}{\xi} + 1; \frac{-\lambda}{\xi})}. \qquad (3.12)
$$

It is known that

$$\frac{{}_1F_1\left(\frac{\mu_2}{\xi}+k+1;\frac{rs+\theta+\mu_v}{\xi}+k+1;\frac{-\lambda}{\xi}\right)}{\prod\limits_{i=1}^{k}\left(\frac{rs+\theta+\mu_2}{\xi}+i\right)}=\xi\sum_{n=0}^{\infty}\prod_{i=1}^{n}\left(\frac{\mu_2}{\xi}+k+i\right)\frac{(-\lambda)^n}{\xi^n n!}$$

$$\sum_{i=1}^{k+n}\frac{(-1)^{i-1}}{(i-1)!(n+k-1)!(rs+\theta+\mu_2+i\xi)},\quad(3.13)$$

and

$${}_1F_1\left(\frac{\mu_2}{\xi}+1;\frac{rs+\theta+\mu_v}{\xi}+1;\frac{-\lambda}{\xi}\right)=\sum_{n=0}^{\infty}(\lambda)^n\hat{a}_n(s),$$

where

$$\hat{a}_n(s)=\frac{\prod\limits_{i=1}^{n}\left(\frac{\mu_2}{\xi}+i\right)(-1)^n}{n!}\sum_{i=1}^{n}\frac{(-1)^{i-1}}{\xi^{n-1}(i-1)!(n-i)!}\frac{1}{(rs+\theta+\mu_2+i\xi)}.$$

Hence

$$\left[{}_1F_1\left(\frac{\mu_2}{\xi}+1;\frac{rs+\theta+\mu_v}{\xi}+1;\frac{-\lambda}{\xi}\right)\right]^{-1}=\sum_{n=0}^{\infty}(\lambda)^n\hat{b}_n(s)\qquad(3.14)$$

where $\hat{b}_0(s)=1$ and for $n=1,2,3,\cdots$

$$\hat{b}_n(s)=\begin{vmatrix}\hat{a}_1(s) & 1 & \cdots & \cdots & \cdots \\ \hat{a}_2(s) & \hat{a}_1(s) & 1 & \cdots & \cdots \\ \hat{a}_3(s) & \hat{a}_2(s) & \hat{a}_3(s) & 1 & \cdots \\ \vdots & \vdots & \vdots & \vdots & \cdots \\ \hat{a}_{n-1}(s) & \hat{a}_{n-2}(s) & \hat{a}_{n-2}(s) & \cdots \hat{a}_1(s) & 1 \\ \hat{a}_n(s) & \hat{a}_{n-1}(s) & \hat{a}_{n-2}(s) & \cdots \hat{a}_2(s) & \hat{a}_1(s)\end{vmatrix}$$

$$=\sum_{i=1}^{n}(-1)^{i-1}\hat{a}_i(s)\hat{b}_{k-i}(s).$$

Inversion of above equation yields

$$b_n(u)=\sum_{i=1}^{n}(-1)^{i-1}a_i(u)*b_{k-i}(u),$$

where

$$a_n(u)=\frac{\prod\limits_{i=1}^{n}\left(\frac{\mu_2}{\xi}+i\right)(-1)^n}{n!}\sum_{i=1}^{n}\frac{(-1)^{i-1}}{r\xi^{n-1}(i-1)!(n-i)!}e^{-\left(\frac{\theta+\mu_2+i\xi}{r}\right)u}.$$

Substituting the Eqs. (3.13) and (3.14) in Eq. (3.12), we get

$$\hat{\Psi}_k(s) = \xi \sum_{n=0}^{\infty} \prod_{i=1}^{n} \left(\frac{\mu_2}{\xi} + k + i \right) \frac{(-\lambda)^n}{\xi^n n!} \sum_{i=1}^{k+n} \frac{(-1)^{i-1}}{(i-1)!(n+k-1)!(rs+\theta+\mu_2+i\xi)}$$

$$\sum_{n=0}^{\infty} (\lambda)^n \hat{b}_n(s).$$

Inversion of above equation gives

$$\Psi_k(u) = \xi \sum_{n=0}^{\infty} \prod_{i=1}^{n} \left(\frac{\mu_2}{\xi} + k + i \right) \frac{(-\lambda)^n}{r\xi^n n!} \sum_{i=1}^{k+n} \frac{(-1)^{i-1}}{(i-1)!(n+k-1)!} e^{-\left(\frac{\theta+\mu_2+i\xi}{r}\right)u}$$

$$* \sum_{n=0}^{\infty} (\lambda)^n b_n(u).$$

3.3   Evaluation of $F_{k,1}(u)$

Define the generating function

$$G(z,u) = \sum_{k=1}^{\infty} F_{k,1}(u) z^k.$$

By standard methods, the system of difference-differential equations represented by Eqs. (2.4) and (2.5) leads to a linear differential equation and integrating, we get

$$G(z,u) = \frac{\theta}{r} \int_0^u \sum_{m=1}^{\infty} F_{m,0}(y) z^m exp\left[-\left(\frac{\lambda+\mu_1}{r} \right)(u-y) \right] exp\left[\left(\frac{\lambda z}{r} + \frac{\mu_1}{rz} \right)(u-y) \right] dy$$

$$- \frac{\mu_1}{r} \int_0^u F_{1,1}(y) exp\left[-\left(\frac{\lambda+\mu_1}{r} \right)(u-y) \right] exp\left[\left(\frac{\lambda z}{r} + \frac{\mu_1}{rz} \right)(u-y) \right] dy$$

$$\tag{3.15}$$

It is well known that if $\alpha_1 = 2\frac{\sqrt{\lambda\mu_1}}{r}$ and $\beta = \sqrt{\frac{\lambda}{\mu_1}}$, then

$$exp\left[\left(\frac{\lambda z}{r} + \frac{\mu_1}{rz} \right)u \right] = \sum_{k=-\infty}^{\infty} (\beta z)^k I_k(\alpha_1 u).$$

Comparing the coefficients of z^k on both sides of Eq. (3.15), for $k = 1, 2 \cdots$ yields

$$F_{k,1}(u) = \frac{\theta}{r} \int_0^u \sum_{m=1}^{\infty} F_{m,0}(y) \beta^{k-m} I_{k-m}(\alpha_1(u-y)) exp\left[-\left(\frac{\lambda+\mu_1}{r} \right)(u-y) \right] dy$$

$$- \frac{\mu_1}{r} \int_0^u F_{1,1}(y) \beta^k I_k(\alpha_1(u-y)) exp\left[-\left(\frac{\lambda+\mu_1}{r} \right)(u-y) \right] dy. \tag{3.16}$$

The above equation holds for $k = -1, -2, -3, \cdots$ with the left-hand side replaced by zero.

Using $I_{-k}(\alpha_1(u - y)) = I_k(\alpha_1(u - y))$ for $k = 1, 2, 3, \cdots$ yields

$$0 = \frac{\theta}{r} \int_0^u \sum_{m=1}^{\infty} F_{m,0}(y)\beta^{-k-m} I_{k+m}(\alpha_1(u - y)) exp\left[-\left(\frac{\lambda + \mu_1}{r}\right)(u - y)\right] dy$$
$$+ \frac{\mu_1}{r} \int_0^u F_{1,1}(y)\beta^{-k} I_k(\alpha_1(u - y)) e^{\left[-\left(\frac{\lambda + \mu_1}{r}\right)(u-y)\right]} dy. \tag{3.17}$$

Summing the Eqs. (3.16) and (3.17) leads to

$$F_{k,1}(u) = \frac{\theta}{r} \int_0^u e^{\left[-\left(\frac{\lambda + \mu_1}{r}\right)(u-y)\right]}$$
$$\sum_{m=1}^{\infty} \beta^{k-m} F_{m,0}(y)[I_{k-m}(\alpha_1(u - y)) - I_{k+m}(\alpha_1(u - y))] dy, \tag{3.18}$$

for $k = 1, 2, 3, \cdots$. Thus, $F_{k,1}(u)$ is expressed in terms of $F_{k,0}(u)$ for $k = 1, 2, 3, \cdots$. It is seen that Eq. (3.11) gives $F_{k,0}(u)$ in terms of $F_{0,0}(u)$. It still remains to determine $F_{0,0}(u)$.

3.4   Evaluation of $F_{0,0}(u)$

Taking Laplace transform of Eq. (2.2) yields

$$(sr_0 + \lambda)\hat{F}_{0,0}(s) = ar_0 + \mu_1 \hat{F}_{1,1}(s) + (\mu_2 + \xi)\hat{F}_{1,0}(s). \tag{3.19}$$

From Eq. (3.18) for $k = 1$, we get

$$F_{1,1}(u) = \frac{\theta}{r} \int_0^u exp\left[-\left(\frac{\lambda + \mu_1}{r}\right)(u - y)\right]$$
$$\sum_{m=1}^{\infty} \beta^{1-m} F_{m,0}(y)[I_{m-1}(\alpha_1(u - y)) - I_{m+1}(\alpha_1(u - y))] dy,$$

and its Laplace transform is given by

$$\hat{F}_{1,1}(s) = \frac{\theta}{\mu_1} \sum_{m=1}^{\infty} \left(\frac{\frac{\lambda}{r\beta}}{s + \frac{\lambda}{r} + \frac{\theta}{r}}\right)^m \left(\frac{p_1 - \sqrt{p_1^2 - \alpha_1^2}}{\alpha_1}\right)^m \hat{F}_{0,0}(s), \tag{3.20}$$

where $p_1 = s + \frac{\lambda + \mu_1}{r}$. Also, from Eq. (3.10) for $k = 1$, we get

$$\hat{F}_{1,0}(s) = \hat{\Psi}_1(s)\hat{F}_{0,0}(s). \tag{3.21}$$

Substituting Eqs. (3.20) and (3.21) in Eq. (3.19), after some algebraic calculations leads to

$$\hat{F}_{0,0}(s) = a \sum_{j=0}^{\infty} \frac{1}{r_0^j (s + \frac{\lambda}{r_0})^{j+1}} \left[\theta \sum_{m=0}^{\infty} \left(\frac{\frac{\lambda}{r\beta}}{s + \frac{\lambda}{r} + \frac{\theta}{r}} \right)^{m+1} \left(\frac{p_1 - \sqrt{p_1^2 - \alpha_1^2}}{\alpha_1} \right)^{m+1} + (\mu_2 + \xi) \hat{\Psi}_1(s) \right]^j .$$

(3.22)

Inversion of the above equation yields

$$F_{0,0}(u) = a \sum_{j=0}^{\infty} \left(\frac{1}{r_0} \right)^j \frac{u^j}{j!} exp \left[-\left(\frac{\lambda}{r_0} \right) u \right] *$$

$$\left[\theta \sum_{m=0}^{\infty} \left(\frac{\lambda}{r\beta} \right)^{m+1} \frac{u^m}{m!} exp \left[-\left(\frac{\lambda}{r} + \frac{\theta}{r} \right) u \right] * \frac{(m+1)I_{m+1}(\alpha_1 u)}{u} + (\mu_2 + \xi)\Psi_1(u) \right]^{*j} .$$ (3.23)

Hence all the joint steady state probabilities of the fluid queueing model are explicitly determined in terms of modified Bessel's function of the first kind using continued fraction and generating function methodologies.

Remark

When the impatience parameter $\xi = 0$, the model under consideration reduces to a fluid queue driven by an *M/M/1* queue subject to working vacation.

4   Conclusion

This paper studies a fluid model driven by an *M/M/1* queue subject to working vacation and customer impatience. The study of such models provide greater flexibility to the design and control of input and output rates of fluid flow thereby adapting the fluid models to wider application background. The governing system of infinite differential difference equations are explicitly solved to obtain explicit analytical expressions for the joint system state probabilities of the state of the background queueing model and the content of the buffer. The various state probabilities during vacation state of the server are solved using continued fraction together with Laplace transformation and the corresponding state during the busy period are solved using generating function methodology. Most of the existing results in the literature pertaining to fluid models has presented the solution to the buffer content distribution in the Laplace domain. However, closed form analytical solutions helps to gain a deeper insight into the model and other related performance measures.

References

1. Gao, S., Wang, J.: Discrete time $Geo^X/G/1$ retrial queue with general retrial times, working vacations and vacation interruption. Qual. Technol. Quant. Manage. **10**(4), 495–512 (2013)
2. Gao, S., Yin, C.: Discrete-time $Geo^X/G/1$ queue with geometrically working vacations and vacation interruption. Qual. Technol. Quant. Manage. **10**(4), 423–442 (2013)
3. Huo, Z., Jin, S., Tian, N.: Performance analysis and evaluation for connection-oriented networks based on discrete time vacation queueing model. Qual. Technol. Quant. Manage. **5**(1), 51–62 (2008)
4. Latouche, G., Taylor, P.G.: A stochastic fluid model for an ad hoc mobile network. Queueing Syst. **63**, 109–129 (2009)
5. Mao, B., Wang, F., Tian, N.: Fluid model driven by an $M/M/1$ queue with multiple exponential vacations and N-policy. J. Appl. Math. Comput. **38**, 119–131 (2012)
6. Mitra, D.: Stochastic theory of a fluid model of producers and consumers couple by a buffer. Adv. Appl. Probab. **20**, 646–676 (1988)
7. Narayanan, C.V., Deepak, T.G., Krishnamoorthy, A., Krishnakumar., B.: On an (s, S) inventory policy with service time, vacation to server and correlated lead time. Qual. Technol. Quant. Manage. **5**(2), 129–143 (2008)
8. Jain, M., Upadhyaya, S.: Threshold N-policy for degraded machining system with multiple type spares and multiple vacations. Qual. Technol. Quant. Manage. **6**(2), 185–203 (2009)
9. Parthasarathy, P.R., Vijayashree, K.V., Lenin, R.B.: An M/M/1 driven fluid queue-continued fraction approach. Queueing Syst. **42**, 189–199 (2002)
10. Ammar, S.I.: Analysis of an $M/M/1$ driven fluid queue with multiple exponential vacations. Appl. Math. Comput. **227**, 329–334 (2014)
11. Vijayashree, K.V., Anjuka, A.: Stationary analysis of an $M/M/1$ driven fluid queue subject to catastrophes and subsequent repair. IAENG Int. J. Appl. Math. **43**(4), 238–241 (2013)
12. Vijaya Lakxmi, P., Goswami, V., Jyothsna, K.: Analysis of discrete-time single server queue with balking and multiple working vacations. Qual. Technol. Quant. Manage. **10**(4), 443–456 (2013)
13. Xu, X., Geng, J., Liu, M., Guo, H.: Stationary analysis for the fluid model driven by the $M/M/c$ working vacation queue. J. Math. Anal. Appl. **403**(2), 423–433 (2013)
14. Xu, X., Zhao, Y., Geng, J., Jin, S.: Analysis for the fluid model driven by an $M/PH/1$ queue. J. Inf. Comput. Sci. **10**(11), 3489–3496 (2013)
15. Yue, D., Yue, W., Xu, G.: Analysis of customers impatience in an $M/M/1$ queue with working vacation. J. Industr. Manage. Optim. **8**(4), 895–908 (2012)
16. Wang, F., Mao, B., Tian, N.: Fluid model driven by an $M/M/1$ queue with multiple exponential vacation. In: The 2nd International Conference on Advanced Computer Control, vol. 3, pp. 112–115 (2010)

Hyers-Ulam Stability of Linear Differential Equations

R. Murali[(⊠)] and A. Ponmana Selvan

PG and Research Department of Mathematics, Sacred Heart College
(Autonomous), Tirupattur, Vellore 635 601, Tamil Nadu, India
shcrmurali@yahoo.co.in, selvaharry@yahoo.com

Abstract. In this paper, we investigate the approximate solution of the homogeneous and non-homogeneous linear differential equations of second order and nth order, where n is even, in the sense of Hyers-Ulam. Also, some illustrative examples are given.

Keywords: Hyers-Ulam stability · Linear differential equations
Approximate solution · Homogeneous and non-homogeneous

1 Introduction

We say that the functional equation is stable, if for every approximate solution there exists an exact solution near to it. The study of stability problem for various functional equations originated from a famous talk of Ulam [20]. In 1940, Ulam [20] posed a problem concerning the stability of functional equations: "Give Conditions in order for a linear function near an approximately linear function to exist."

Since then, this question has attracted the attention of many researchers. The first solution to this question was given by Hyers [11] in 1941. He made a significant breakthrough, when he gave an affirmative answer to the Ulam's problem for additive functions defined on Banach Spaces. Thereafter, the result by Hyers [11] was generalized by Rassias [17], Aoki [2] and Bourgin [4]. After that, many Mathematicians have extended the Ulam's problem to other functional equations and generalized the Hyers results in various directions (see [2, 4–8, 22, 23, 25, 26]).

Definition of Hyers-Ulam stability has applicable significance. Because it means that if one studies a Hyers-Ulam stable system, one need not reach the exact solution. (Which usually is quite difficult or time consuming). Hyers-Ulam stable system is quite useful in many applications. For Example, Fluid Dynamics, numerical analysis, optimization, biology, statistics, non-linear analysis [16] and economics etc., where finding the exact solution is quite difficult. Also, for every applications of finding the solution of the differential equation, there are corresponding applications for finding the approximate solution or Hyers-Ulam stability of the differential equations.

© Springer Nature Singapore Pte Ltd. 2018
G. Ganapathi et al. (Eds.): ICC3 2017, CCIS 844, pp. 183–192, 2018.
https://doi.org/10.1007/978-981-13-0716-4_15

The theory of stability is an important branch of the qualitative theory of differential equations. The generalization of Ulam's problem was recently proposed by replacing functional equations with differential equations: The differential equation $\phi(x,x',x'',\ldots,x^{(n)}) = 0$ has the Hyers-Ulam stability if for a given ϵ and a function x such that $|\phi(x, x', x'', \ldots, x^{(n)})| \leq \epsilon$, and there exists a solution xaof the differential equation such that $|x(t) - x_a(t)| \leq K(\epsilon)$ and $\lim_{n\to\infty} K(\epsilon) = 0$.

Oblaza seems to be the first author who investigated the Hyers-Ulam stability of linear differential equation [14, 15]. Thereafter, Alsina and Ger [1] published their papers, which handled the Hyers-Ulam stability of the linear differential equation $y'(t) = y(t)$. The result obtained by Alsina and Ger was generalized by Takahasi et al. [19] to the case of the complex Banach Space valued differential equation $y'(t) = \lambda y(t)$, (see also [18, 21]).

These days, the Hyers-Ulam stability of ordinary differential equations has been investigated (see [3, 9, 10, 12, 13, 18, 21, 24, 27, 28]) and the investigation is ongoing.

The foremost aim of this paper is to prove the Hyers-Ulam stability of the homogeneous and non-homogeneous second order linear differential equation of the form

$$x''(t) - p(t)x'(t) + x(t) = 0 \tag{1}$$

and

$$x''(t) - p(t)x'(t) + x(t) = \phi(t) \tag{2}$$

where, xt be the twice continuously differentiable function and the nth order homogeneous and non-homogeneous linear differential equation of the form

$$x^{(n)}(t) - p(t)x^{(n-1)}(t) + x^{(n-2)}(t) - \cdots + x''(t) - p(t)x(t) + x(t) = 0 \tag{3}$$

and

$$x^{(n)}(t) - p(t)x^{(n-1)}(t) + x^{(n-2)}(t) - \cdots + x''(t) - p(t)x(t) + x(t) = \phi(t) \tag{4}$$

where, $x \in C^n[a,b], p, \phi \in C[a,b]$ for all $a, b \in \mathbb{R}$ with $-\infty < a < b < \infty$ and n is even.

2   Hyers-Ulam Stability of (1) and (2)

In this section, we would like to prove the Hyers-Ulam stability of the homogeneous and non-homogeneous second order linear differential Eqs. (1) and (2). First, we prove the Hyers-Ulam stability of (1).

Theorem 1. Let $x(t)$ be twice continuously differentiable function on [a, b], then the differential Eq. (1) has the Hyers-Ulam stability.

Proof. For every $\epsilon > 0$, there exists $x \in C^2[a, b]$, satisfies the following inequality

$$|x''(t) - p(t)x'(t) + x(t)| \leq \epsilon \tag{5}$$

for all $t \in [a, b]$. Now, define

$$q(t) = x'(t), P(t) = \exp\left(\int_a^t p(s)ds\right) z(t) = \frac{q(b)}{P(b)} P(t) \text{ for all } t \in [a, b].$$

Then we have, $z'(t) = p(t)z(t)$. Also, we have

$$|q'(t) - p(t)q(t)| = |x''(t) - p(t)x'(t)| \leq |x''(t) - p(t)x'(t) + x(t)| \leq \epsilon.$$

Now, $|z(t) - q(t)| \leq P(t) \int_t^b \frac{1}{P(s)} |q'(s) - q(s)p(s)|ds \leq P(t)\left(\int_t^b \frac{1}{P(s)} ds\right)\epsilon,$

for all $t \in [a, b]$. Since, $p \in C[a, b]$, there exists a constant m, M such that $m \leq p(t) \leq M$ and so $e^{m(t-a)} \leq P(t) \leq e^{M(t-a)}$ for all $t \in [a, b]$. Hence we arrive

$$|z(t) - q(t)| \leq \epsilon e^{M(t-a)} \int_t^b e^{-mt} dt.$$

So,

$$|z(t) - q(t)| \leq \begin{cases} e^{M(b-a)} \cdot \frac{1 - e^{m(b-a)}}{m} \epsilon & \text{if } M \geq 0, m \neq 0; \\ \frac{1 - e^{m(b-a)}}{m} \epsilon & \text{if } M < 0 \\ (b-a)e^{M(b-a)} \epsilon & \text{if } m = 0, \end{cases} \tag{6}$$

for all $t \in [a, b]$. Let us consider $h(t) = x(t)$ and

$$k(t) = e^{i(t-b)}h(b) - e^{it} \int_t^b z(s)e^{-is}ds,$$

for all $t \in [a, b]$. Hence,

$$k'(t) - ik(t) = z(t) \tag{7}$$

$$|h'(t) - ih(t) - z(t)| < |z(t) - q(t)| \tag{8}$$

for all $t \in [a, b]$. Now, we define,

$$y(t) = x(b)e^{-i(t-b)} - e^{-it} \int_t^b k(s)e^{is}ds,$$

for all $t \in [a, b]$. Obviously, we have $y \in C^2[a, b]$ and $k(t) = x(t)$. Therefore, from (7), we obtain

$$z(t) = y'(t) + y(t), \tag{9}$$

for all $t \in [a, b]$. Also, we arrive at

$$|k(t) - h(t)| = \left| \int_t^b (e^{-is} h(s))' ds - \int_t^b z(s) e^{-is} ds \right| \le \int_t^b |h'(s) - ih(s) - z(s)| ds,$$

for all $t \in [a, b]$. Hence from (6) and (8), we have

$$|k(t) - h(t)| \le \begin{cases} (b-a) e^{M(b-a)} \cdot \frac{1 - e^{m(a-b)}}{m} \epsilon & \text{if } M \ge 0, m \ne 0; \\ (b-a) \frac{1 - e^{m(a-b)}}{m} \epsilon & \text{if } M < 0 \\ (b-a)^2 e^{M(b-a)} \epsilon & \text{if } m = 0, \end{cases} \tag{10}$$

for all $t \in [a, b]$. Then from (5) and (9) we have $y''(t) - p(t) y'(t) + y(t) = 0$.

$$\therefore |x(t) - y(t)| \le \left| \int_t^b k(s) e^{is} ds - \int_t^b [x'(s) + ix(s)] e^{is} ds \right| \le \int_t^b |k(s) - h(s)| ds,$$

for all $t \in [a, b]$. Then we have,

$$|x(t) - y(t)| \le \begin{cases} (b-a)^2 e^{M(b-a)} \cdot \frac{1 - e^{m(a-b)}}{m} \epsilon & \text{if } M \ge 0, m \ne 0; \\ (b-a)^2 \frac{1 - e^{m(a-b)}}{m} \epsilon & \text{if } M < 0; \\ (b-a)^3 e^{M(b-a)} \epsilon & \text{if } m = 0, \end{cases}$$

for all $t \in [a, b]$, this completes the proof.

Now, we are going to prove the Hyers-Ulam stability of the non-homogeneous linear differential Eq. (2).

Theorem 2. Let $x(t)$ be twice continuously differentiable function on [a, b], then the differential Eq. (2) has the Hyers-Ulam stability.

Proof. For every $\epsilon > 0$, there exists $x \in C^2[a, b]$, satisfies the following inequality

$$|x'(t) - p(t) x'(t) + x(t) - \phi(t)| \le \epsilon$$

for all $t \in [a, b]$. Now, define $q(t) = x'(t)$,

$$P(t) = \exp\left(\int_a^t p(s) ds \right), \qquad z(t) = \frac{q(b)}{P(b)} P(t) - \int_a^t \frac{\phi(s)}{P(s)} ds \; \forall t \in [a, b].$$

Then we have, $z'(t) = p(t) z(t) + \phi(t)$. Also, we have

$$|q'(t) - p(t)q(t) - \phi(t)| \le |x''(t) - p(t)x'(t) + x(t) - \phi(t)| \le \epsilon.$$

Now,

$$|z(t) - q(t)| \le P(t) \int_t^b \frac{1}{P(s)} |q'(s) - q(s)p(s) - \phi(t)| ds$$

$$\le P(t) \left(\int_t^b \frac{1}{P(s)} ds \right) \epsilon, \text{ for all } t \in [a, b].$$

By applying the similar argument of the proof of Theorem 1, we can easily prove this Theorem.

3 Hyers-Ulam Stability of (3) and (4)

Now, in the following theorems, we establish the Hyers-Ulam stability of the nth order homogeneous and non-homogeneous linear differential Eqs. (3) and (4). First, we are going to prove the Hyers-Ulam stability of the linear differential Eq. (3).

Theorem 3. Let $x(t)$ be n times continuously differential function on $[a, b]$, where n is even, then the differential Eq. (3) has the Hyers-Ulam stability.

Proof. For every $\epsilon > 0$, there exists $x \in C^n[a, b]$, satisfies the following inequality

$$\left| x^{(n)}(t) - p(t)x^{(n-1)}(t) + x^{(n-2)}(t) + \ldots + x''(t) - p(t)x'(t) + x(t) \right| \le \epsilon, \qquad (11)$$

for all $t \in [a, b]$. Now, define $q(t) = x^{n-1}(t) + x^{n-3}(t) + \ldots + x'(t)$,

$$P(t) = \exp \left(\int_a^t p(s)ds \right) \quad and \quad z(t) = \frac{q(b)}{P(b)} P(t),$$

for all $t \in [a, b]$. Then we have, $z'(t) = p(t)z(t)$. Also, we have

$$|q'(t) - p(t)q(t)| = \left| x^{(n)}(t) - p(t)x^{(n-1)}(t) + \ldots - p(t)x'(t) + x(t) \right| \le \epsilon.$$

Now, $|z(t) - q(t)| \le P(t) \int_t^b \frac{1}{P(s)} |q'(s) - q(s)p(s)| ds \le P(t) \left(\int_t^b \frac{1}{P(s)} ds \right) \epsilon$,

for all $t \in [a, b]$. Since, $p \in C[a, b]$, there exists a constant m, M such that $m \le p(t) \le M$ and so $e^{m(t-a)} \le P(t) \le e^{M(t-a)}$ for all $t \in [a, b]$. Hence we arrive

$$|z(t) - q(t)| \le \epsilon e^{M(t-a)} \int_t^b e^{-mt} dt \le \begin{cases} e^{M(b-a)} \cdot \frac{1-e^{m(a-b)}}{m} \epsilon & \text{if } M \ge 0, m \ne 0; \\ \frac{1-e^{m(a-b)}}{m} \epsilon & \text{if } M < 0 \\ (b-a)e^{M(b-a)} \epsilon & \text{if } m = 0, \end{cases} \quad \forall t \in [a, b].$$

$$(12)$$

Let us consider $h(t) = x^{(n-2)}(t) + ix^{n-3}(t) + \ldots + x''(t) + ix'(t)$ and

$$k(t) = e^{i(t-b)}h(b) - e^{it}\int_t^b z(s)e^{-is}ds,$$

for all $t \in [a, b]$. Hence,

$$k'(t) - ik(t) = z(t) \tag{13}$$

$$|h'(t) - ih(t) - z(t)| < |z(t) - q(t)| \tag{14}$$

for all $t \in [a, b]$. Now, we define,

$$y(t) = x(b)e^{-i(t-b)} - e^{-it}\int_t^b k(s)e^{is}ds, \qquad \forall t \in [a, b].$$

Obviously, we have $y \in C^n[a, b]$ and

$$k(t) = x^{(n-2)}(t) + ix^{n-3}(t) + \ldots + x''(t) + ix'^{(t)}.$$

Therefore, from (13) we obtain

$$z(t) = y^{(n-1)}(t) + y^{n-3}(t) + \ldots + y'''(t) + y'(t), \tag{15}$$

for all $t \in [a, b]$. Also, we arrive at

$$|k(t) - h(t)| = \left| e^{ib}h(b) - e^{-it}h(t) - \int_t^b z(s)e^{-is}ds \right|$$

$$\leq \int_t^b |h'(s) - ih(s) - z(s)|ds,$$

for all $t \in [a, b]$. Hence from (12) and (14), we have

$$|k(t) - h(t)| \leq \begin{cases} (b-a)e^{M(b-a)} \cdot \frac{1-e^{m(a-b)}}{m}\epsilon & \text{if } M \geq 0, m \neq 0; \\ (b-a)\frac{1-e^{m(a-b)}}{m}\epsilon & \text{if } M < 0 \\ (b-a)^2 e^{M(b-a)}\epsilon & \text{if } m = 0, \end{cases} \tag{16}$$

for all $t \in [a, b]$. Then from (11) and (15), we have

$$x^{(n)}(t) - p(t)x^{(n-1)}(t) + x^{(n-2)}(t) + \ldots + x''(t) - p(t)x'(t) + x(t) = 0.$$

Now, $|x(t) - y(t)| \leq \left| \int_t^b k(s)e^{is}ds - \int_t^b [x'(s) + ix(s)]e^{is}ds \right| \leq \int_t^b |k(s) - h(s)|ds$, for all $t \in [a, b]$. Then we have,

$$|x(t) - y(t)| \leq \begin{cases} (b-a)^2 e^{M(b-a)} \cdot \frac{1-e^{m(a-b)}}{m} \epsilon & \text{if } M \geq 0, m \neq 0; \\ (b-a)^2 \frac{1-e^{m(a-b)}}{m} \epsilon & \text{if } M < 0 \\ (b-a)^3 e^{M(b-a)} \epsilon & \text{if } m = 0, \end{cases}$$

for all $t \in [a,b]$. this completes the proof.

Finally, we are going to prove the Hyers-Ulam stability of the non-homogeneous differential Eq. (4), by applying the similar argument of the proof of Theorem 3. With this we can prove the following Theorem.

Theorem 4. Let $x(t)$ be n times continuously differential function on $[a, b]$, where n is even, then the differential Eq. (4) has the Hyers-Ulam stability.

Proof. For every $\epsilon > 0$, there exists $x \in C^n[a,b]$, satisfies the following inequality

$$\left| x^{(n)}(t) - p(t)x^{(n-1)}(t) + x^{(n-2)}(t) + \ldots + x''(t) - p(t)x'(t) + x(t) - \phi(t) \right| \leq \epsilon,$$

for all $t \in [a,b]$. Now, define $q(t) = x^{n-1}(t) + x^{n-3}(t) + \ldots + x'(t)$,

$$P(t) = \exp\left(\int_a^t p(s)ds \right) \quad \text{and} \quad z(t) = \frac{q(b)}{P(b)}P(t) - P(t)\int_a^t \frac{\phi(s)}{P(s)}ds$$

for all $t \in [a,b]$. Then we have, $z'(t) = p(t)z(t) + \phi(t)$. Also, we have

$$|q'(t) - p(t)q(t) - \phi(t)|$$
$$= \left| x^{(n)}(t) - p(t)x^{(n-1)}(t) + \ldots x''(t) - p(t)x'(t) + x(t) - \phi(t) \right| \leq \epsilon.$$

Now,

$$|z(t) - q(t)| \leq P(t) \int_t^b \frac{1}{P(s)} |q'(s) - q(s)p(s) - \phi(t)|ds$$

$$\leq P(t)\left(\int_t^b \frac{1}{P(s)}ds \right)\epsilon, \quad \forall t \in [a,b].$$

Hence, by a similar method as we applied to the proof of Theorem 3, one can prove this Theorem easily.

Now, we are going to illustrate the Theorems 1 and 2 by the following examples.

Example 5. Let $I = [a, b]$ be an interval with a and b are real numbers. Suppose that the mapping x: $I \to R$ be twice continuous differentiable function. Let $p(t) \in C(I)$ is defined as $p(t) =: K$, where K is positive constant and for all t in I. Then the differential equation $x''(t) - p(t)x'(t) + x(t) = 0$, has the Hyers-Ulam stability.

Solution: Let $\epsilon > 0$, and x(t) be a twice continuously differentiable function satisfying the inequality, $|x''(t) - p(t)x'(t) + x(t)| \leq \epsilon$ for all $t \in I$. Let $q(t) = x'(t)$ and $P(t) = e^{\int_a^t p(s)ds} = e^{K(t-a)}$.

Now, we have z(t) $= z(t) = \frac{q(b)}{e^{K(b-a)}} e^{K(t-a)}$, for all $t \in I$, with $z'(t) = K z(t)$. Also, we have that $|q'(t) - Kq(t)| \leq \epsilon$, with

$$|z(t) - q(t)| \leq \frac{e^{K(t-a)}}{K} \left(e^{K(b-a)} - e^{K(t-a)} \right) \epsilon,$$

for all $t \in I$. Since, $p \in C[a, b]$, there exists a constant m, M such that $m \leq p(t) \leq M$ and so $e^{m(t-a)} \leq P(t) \leq e^{M(t-a)}$ for all $t \in [a, b]$. Hence, by using Theorem 1, we arrive the Hyers-Ulam stability of the differential equation (1).

Example 6. Let I = [a, b] be an interval with a and b are real numbers. Suppose that the mapping x: I $\to$ R be twice continuous differentiable function. Let $p(t) \in C(I)$ is defined as $p(t) =: K$ where $K > 0$ is a constant and for all t in I. Then the differential equation $x''(t) - p(t)x'(t) + x(t) = \alpha(t)$, has the Hyers-Ulam stability.

Solution: Let $\epsilon > 0$, and $x(t)$ be a twice continuous differentiable function satisfying the inequality, $|x''(t) - p(t)x'(t) + x(t) - \alpha(t)| \leq \epsilon$ for all $t \in I$. Let $q(t) = x'(t)$ and

$$P(t) = e^{\int_a^t p(s)ds} = e^{K(t-a)} \text{ and } z(t) = \frac{q(b)}{e^{K(b-a)}} e^{K(t-a)} - e^{K(t-a)} \int_a^t \frac{\alpha(s)}{e^{K(s-a)}} ds$$

for all $t \in I$, with $z'(t) = Kz(t) + \alpha(t)$. Also, we have that $|q'(t) - Kq(t) - \alpha(t)| \leq \epsilon$ with

$$|z(t) - q(t)| \leq \frac{e^{K(t-a)}}{K} \left(e^{K(b-a)} - e^{K(t-a)} \right) \epsilon$$

for all $t \in I$. Since, $p \in C[a, b]$, there exists a constant m, M such that $m \leq p(t) \leq M$ and so $e^{m(t-a)} \leq P(t) \leq e^{M(t-a)}$ for all $t \in [a, b]$. Hence, by using Theorem 2, we arrive the Hyers-Ulam stability of the differential equation (2).

4 Conclusion

Here, we have obtained the sufficient criteria for the Hyers-Ulam stability of homogeneous and non-homogeneous linear differential equations of second order and nth order. It is very useful to the readers to find the approximate solution of the differential equations of second order and nth order.

Acknowledgements. The authors express their sincere gratitude to the editors and anonymous reviewers for the careful reading of the original manuscript, for the useful comments that helped to improve the presentation of the results and for accentuating the important details.

References

1. Alsina, C., Ger, R.: On some inequalities and stability results related to the exponential function. J. Inequal. Appl. **2**, 373–380 (1998)
2. Aoki, T.: On the linear transformation in Banach space. J. Math. Soc. Jpn. **2**, 64–66 (1950)
3. Alqifiary, Q.H., Jung, S.-M.: Hyers-Ulam stability of second order linear differential equations with boundary conditions. SYLWAN **158**(5), 289–301 (2014)
4. Bourgin, D.G.: Classes of transformation and bordering transformations. Bull. Amer. Math. Soc. **57**, 223–237 (1951)
5. Brillouet-Belluot, N., Brzdek, J., Cieplinski, K.: On some recent developments in Ulam's type stability. Abs. Appl. Anal. **2012**, 41 (2012). Article ID 716936
6. Brzdek, J., Popa, D., Xu, B.: Remarks on stability of linear recurrence of higher order. Appl. Math. Lett. **23**, 1459–1463 (2010)
7. Burger, M., Ozawa, N., Thom, A.: On Ulam stability. Israel J. Math. **193**, 109–129 (2013)
8. Chung, J.: Hyers-Ulam stability theorems for Pexiders equations in the space of Schwartz distribution. Arch. Math. **84**, 527–537 (2005)
9. Gavruta, P., Jung, S.-M., Li, Y.: Hyers-Ulam stability of second order linear differential equations with boundary conditions. Elec. J. Differ. Equ. **2011**(80), 1–5 (2011)
10. Huang, J., Li, Y.: Hyers-Ulam stability of linear differential equations. J. Math. Anal. Appl. **426**, 1192–1200 (2015)
11. Hyers, D.H.: On the stability of the linear functional equation. Proc. Nat. Acad. Sci. USA **27**, 222–224 (1941)
12. Li, Y., Shen, Y.: Hyers-Ulam stability of linear differential equations of second order. Appl. Math. Lett. **23**, 306–309 (2010)
13. Modebei, M.I., Olaiya, O.O., Otaide, I.: Generalized Hyers-Ulam stability of second order linear ordinary differential equation with initial and conditions. Adv. Inequal. Appl., 1–7 (2014)
14. Oblaza, M.: Hyers stability of the linear differential equation. Rocknik. Nauk-Dydakt Prace Mat. **13**, 259–270 (1993)
15. Oblaza, M.: Connections between Hyers and Lyapunov stability of the ordinary differential equations. Rocknik. Nauk-Dydakt Prace Mat. **14**, 141–146 (1997)
16. Issac, G., Rassias, Th.M.: Stability of additive mappings: applications to non-linear analysis. Int. J. Math. Math. Sci. **19**, 219–228 (1996)
17. Rassias, Th.M.: On the stability of the linear mappings in Banach Spaces. Proc. Am. Math. Soc. **72**, 297–300 (2010)
18. Rus, I.A.: Ulam stabilities of ordinary differential equations in Banach space. Carpathian J. Math. **26**(1), 103–107 (2010)
19. Takahasi, S., Miura, T., Miyajima, S.: On the Hyers-Ulam stability of the Banach space valued differential equation $y'(t)=\lambda y(t)$. Bull. Korean Math. Soc. **39**, 309–315 (2002)
20. Ulam, S.M.: A Collection of Mathematical Problems. Interscience Publishers, New York (1960)
21. Wang, G., Zhou, M., Sun, L.: Hyers-Ulam stability of linear differential equations of first order. Appl. Math. Lett. **21**, 1024–1028 (2008)
22. Xu, T.: On the stability of Multi-Jensen mapping in β-normed spaces. Appl. Math. Lett. **25** (11), 1866–1870 (2012)
23. Zada, A., Shah, O., Shah, R.: Hyers-Ulam stability of non-autonomous system in terms of boundedness of Cauchy problems. Appl. Math. Comput. **271**, 512–518 (2015)

24. Abdollahpoura, M.R., Aghayaria, R., Rassias, T.: Hyers-Ulam stability of associated Laguerre differential equations in a subclass of analytic functions. J. Math. Anal. Appl. **437**, 605–612 (2016)
25. Nash Jr., J.F., Rassias, T.M. (eds.): Open Problems in Mathematics. Springer, New York (2016). https://doi.org/10.1007/978-3-319-32162-2
26. Kannappan, P.: Functional Equations and Inequalities with Applications. Springer, New York (2009). https://doi.org/10.1007/978-0-387-89492-8
27. Murali, R., Ponmana Selvan, A.: On the generalized Hyers-Ulam stability of linear ordinary differential equations of higher order. Int. J. Pure Appl. Math. **117**(12) 317–326 (2017
28. Murali, R., Ponmana Selvan, A.: Hyers-Ulam-Rassias stability for the linear ordinary differential equation of third order. Kragujevac J. Math. **42**(4) 579–590 (2018)

Cost Analysis of an Unreliable Retrial Queue Subject to Balking, Bernoulli Vacation, Two Types of Service and Starting Failure

D. Arivudainambi$^{(\boxtimes)}$ and M. Gowsalya$^{(\boxtimes)}$

Department of Mathematics, College of Engineering Guindy,
Anna University, Chennai, India
arivu@annauniv.edu, gowsalya.maha89@gmail.com

Abstract. The proposed queueing model deals with a single server retrial queueing subject to balking, Bernoulli vacation, two types of service and starting failure. Single server provides two types of heterogeneous service in which the customer will choose either type 1 service or type 2 service with two different probabilities. The authors assume that, while rendering service to the arriving primary or repeated customers the server may subject to starting failure. After each service completion the server will take vacation subject to Bernoulli vacation policy. For such queueing model the necessary and sufficient condition have been derived. The model have been solved by the use of supplementary variable technique. Cost analysis have been carried out with the cost parameters. To validate the proposed model some performance measures, special cases and sensitivity analysis have been discussed.

Keywords: Retrial queues · Bernoulli vacation · Starting failure
Two types of service · Steady state · Cost analysis · Performance measures

1 Introduction

Queueing system have vast applications in many fields because of its optimization technique. Real life areas like communication centers, banks, service centers, production managements, wireless communication protocols it plays an important role. Motivation comes from the real life problems faced in call centers as in bound centers and out bound centers which is termed as two types of service, maintenance carried termed as vacation and during the start of the service to the customers, the server corresponds to the service facility, facing an unexpected failure termed as starting failure, the customers those who didn't get served immediately would try for their request after some random time termed as retrial attempt or retrial queue. This paper is organized as follows. Mathematical description of the model have been elaborated in Sect. 2. Section 3 deals with the equations governing to our model. The steady state solution have been found in Sect. 4. Performance measures and special cases have been discussed in Sects. 5 and 6. Cost analysis of the model analyzed by the use of MATLAB software in Sect. 7.

© Springer Nature Singapore Pte Ltd. 2018
G. Ganapathi et al. (Eds.): ICC3 2017, CCIS 844, pp. 193–204, 2018.
https://doi.org/10.1007/978-981-13-0716-4_16

2 Literature Survey

This section elaborates the survey carried out for the formulation of the proposed queueing model.

2.1 Survey on Retrial Queues

Earlier queueing system have been characterized in such a way that the arriving jobs (customers) to the system gets served immediately if it finds the server free or else it leaves the queueing system. But in many practical queueing phenomenon this classical model wouldn't be efficient to provide an optimum service. To overcome the absence of the server, during the arrival of jobs(customers) retrial queues have been introduced into queueing system. In retrial queueing system, the arriving jobs(customers) finds the server free gets served immediately or in contrary if it finds the server busy or absence in the service facility the jobs must leave the service area and join into a pool of unsatisfied customers called orbit, and repeat its request for service after some random time. To study the retrial queues different retrial queues have been introduced namely, classical retrial policy studied by Yang and Templeton [18], constant retrial policy which was introduced by Fayolle [7], Artalejo and Gomez-Corral [4] introduced linear retrial policy and general retrial times was done by Kapyrin [8]. Survey on retrial queues done by Artalejo [1] which gave an detailed idea about retrial queues.

2.2 Survey on Bernoulli Vacation

Server absence for some random duration due to server maintenance activity, performing some secondary jobs or automatic programmed interruptions have been faced by the jobs (customers) more often while coming into the service facility, which was named as vacation in queueing terminology. Doshi [6] gave a survey on queueing system with vacations, which gave a clear view about vacations. In this proposed model vacation studied based on Bernoulli vacation policy. Bernoulli vacation policy have been characterized by the phenomenon that after each service completion the server will take a vacation of random length with probability q or remains in the system with probability $1 - q$. This Bernoulli schedule discipline have been introduced by Keilson and Servi [11] in their study on oscillating random walk models for $GI/G/1$ vacation system. Krishnakumar and Arivudainambi [12] have analyzed a $M/G/1$ retrial queue with Bernoulli schedule vacation and general retrial times. Ke et al. [9] made a short survey on recent developments in vacation queueing models. An unreliable retrial queue with delaying repair and general retrial times under Bernoulli vacation schedule have been investigated by Choudhury and Ke [5].

2.3 Survey on Impatient Behavior

To design an optimized queueing model, impatient behavior of the jobs (customers) to be considered as vital. One of the impatient behavior which have been often happens in the queueing system is balking of the jobs (customers), because they are not convinced by immediate service or absence of the server for a random period. During the arrival of the primary jobs (customers) if it finds the server busy or in vacation it balks the system with some probability or joins into the retrial queue (orbit) with some other probability termed as the main characteristics of balking. Ke and Chang [10] have studied a modified vacation policy for $M/G/1$ retrial queue with balking and feedback in which they examined the feedback by Bernoulli feedback policy. Arivudainambi and Godhandaraman [2] have analysed a retrial queueing system with balking, optional service and vacation in which thay have investigated optional service model by means of first essential service and second optional service.

2.4   Survey on Unreliable Server

Providing an efficient service always needs a reliable server, but sometimes there may be a situation to face unreliability of the server due to unexpected failure of the server by facing random breakdown, starting failure, negative customers etc. Unreliable retrial queues as starting failure have been investigated by Krishna Kumar et al. [13] as $M/G/1$ retrial queue with feedback and starting failures. Mokaddis et al. have analysed a feedback retrial queueing system with starting failures and single vacation where they examined the feedback as Bernoulli feedback policy. Sumitha and Udaya Chandrika [16] have analysed a retrial queueing system with starting failure, single vacation and orbital search. Analysis of repairable $M^{[X]}/(G_1, G_2)/1$ - feedback retrial G-queue with balking and starting failures under at most J vacations have investigated by Rajadurai et al. [14].

3   The Mathematical Description

Primary customer arrive to the system according to a Poisson process with arrival rate $\lambda > 0$. During the arrival of the customers if the server is free then he/she immediately get their request fulfilled and in contrary if the customer finds the server busy or in repair the arrived customer obliged to leave the service area and repeat its request for service after some random time. We considered one of the impatience behavior of the customers called balking, in this system if the server is busy the arriving customer either balks the system with probability $1 - b$ or joins the orbit with probability b. The primary and repeated customers are said to be in orbit when they finds the server busy and they retry for service in First Come First Served discipline. Inter-retrial times are governed by an arbitrary probability distribution function $A(x)$ with corresponding density function $a(x)$. Single server provides two types of service that is type 1 service with probability θ_1 and type 2 service with probability θ_2, with probability distribution functions $B_1(x)$ and $B_2(x)$, with corresponding density functions $b_1(x)$ and $b_2(x)$. The primary or repeated customers who were about to get service have the options to choose either type 1 service or type 2 service. In our proposed model the single server have take vacation by means of Bernoulli vacation policy. After each service completion the server will take a vacation with probability q or else the server remains in the system with probability $p(1 - q)$. Vacation time have probability distribution function as $V(x)$ with corresponding density function $v(x)$. The single server may subject to starting failure and then the failed server immediately sent for repair with probability distribution function $R(x)$ and corresponding density function $r(x)$. Inter arrival times, service times, vacation times, repair times are assumed to be mutually independent and $\theta(x)dx$, $\mu_i(x)dx$, $\eta(x)dx$, $v(x)dx$ are the conditional completion rates of retrial, i^{th} type service, vacation and repair given that the elapsed time is x.

$$\mu_i(x) = \frac{b_i(x)}{1 - B_i(x)}, i = 1, 2; \quad \theta(x) = \frac{a(x)}{1 - A(x)}; \quad \eta(x) = \frac{v(x)}{1 - V(x)}; \quad v(x) = \frac{r(x)}{1 - R(x)}$$

The state of the system at time t can be defined as Markov process $\{N(t); t \geq 0\} = \{\xi_0(t), \xi_1(t), \xi_2(t), \xi_3(t), \xi_4(t), t \geq 0)\}$ where $X(t)$ denotes the number of customers in orbit at time t and

$$C(t) = \begin{cases} 0, & \text{if the server is idle} \\ 1, & \text{if the server is busy on Type1 service} \\ 2, & \text{if the server is busy on Type2 service} \\ 3, & \text{if the server is on vacation} \\ 4, & \text{if the server is on repair} \end{cases}$$

If $C(t) = 0$ and $X(t) > 0$, then $\xi_0(t)$ represents the elapsed retrial time at time t, if $C(t) = 1$, then $\xi_1(t)$ represents the elapsed time of the customer being served in type 1 service, if $C(t) = 2$, then $\xi_2(t)$ represents the elapsed time of the customer being served in type 2 service, if $C(t) = 3$, then $\xi_3(t)$ represents the elapsed vacation time at time t and if $C(t) = 4$, then $\xi_4(t)$ represents the elapsed repair time at time t.

3.1 Ergodicity Condition

Let $\{t_n; n \in N\}$ be the sequence of epochs of either service completion times or vacation termination time. The sequence of random vectors $Z_n = \{C(t_n +), X(t_n +)\}$ form a Markov chain which is the embedded Markov chain for our queueing system. Its state space is $S = \{0, 1, 2, 3 \text{ and } 4\} \times N$.

Theorem 1. *The embedded Markov chain* $\{Z_n; n \in N\}$ *is ergodic if and only if* $\alpha\lambda b\theta_1\beta_1^1 + \alpha\lambda b\theta_2\beta_2^1 + \alpha\lambda bqV_1 + \bar{\alpha}(1 + \lambda bR_1) < \bar{A}(\lambda)$.

Proof: It is clear that $\{Z_n; n \in N\}$ is an irreducible and aperiodic Markov chain. To prove the ergodicity we may use Foster's criterion, which states that an irreducible and aperiodic Markov chain is ergodic if there exists a non-negative function $f(j), j \in N$ and $\varepsilon > 0$ such that the mean drift $\chi_j = E[f(z_{n+1}) - f(z_n)|z_n = j]$ is finite for all $j \in N$ and $\chi_j \leq -\varepsilon$ for all $j \in N$, except perhaps for some finite number j. In this case, consider the function $f(j) = j$, then we have

$$\chi_j = \begin{cases} \alpha\lambda b\theta_1\beta_1^1 + \alpha\lambda b\theta_2\beta_2^1 + \alpha\lambda bqV_1 + \bar{\alpha}(1 + \lambda bR_1)) - \bar{A}(\lambda), j = 1, 2, \cdots \\ \alpha\lambda b\theta_1\beta_1^1 + \alpha\lambda b\theta_2\beta_2^1 + \alpha\lambda bqV_1 + \bar{\alpha}(1 + \lambda bR_1)) - 1, j = 0 \end{cases}$$

The inequality $\alpha\lambda b\theta_1\beta_1^1 + \alpha\lambda b\theta_2\beta_2^1 + \alpha\lambda bqV_1 + \bar{\alpha}(1 + \lambda bR_1) < \bar{A}(\lambda)$ is a necessary and sufficient condition for ergodicity. The necessary condition follows from Kaplan's condition as noted in Sennot et al. [15], namely $\chi_j < \infty$ for all $j \geq 0$ and there exists $j_0 \in N$ such that $\chi_j \geq 0$ for $j \geq j_0$. In this case, Kaplan's condition is fulfilled because there exists h such that $r_{ij} = 0$ for $j < i - h$ and $i > 0$, where $R = (r_{ij})$ is the one step transition matrix of $\{Z_n; n \geq 1\}$. Then, the inequality $\alpha\lambda b\theta_1\beta_1^1 + \alpha\lambda b\theta_2\beta_2^1 + \alpha\lambda bqV_1 + \bar{\alpha}(1 + \lambda bR_1) \geq \bar{A}(\lambda)$ implies the non-ergodicity of the Markov chain.

4 Equations Governing the System

For the Markov process $\{N(t); t \geq 0\}$, we define the probability

$$P_0(t) = P\{C(t) = 0, X(t) = 0\}$$

and the probability densities

$$P_n(x,t)dx = P\{C(t) = 0, X(t) = n, x \leq \xi_0(t) \leq x + dx\}, t \geq 0, x \geq 0, n \geq 1$$

$$Q_n^1(x,t)dx = P\{C(t) = 1, X(t) = n, x \leq \xi_1(t) \leq x + dx\}, t \geq 0, x \geq 0, n \geq 0$$

$$Q_n^2(x,t)dx = P\{C(t) = 2, X(t) = n, x \leq \xi_2(t) \leq x + dx\}, t \geq 0, x \geq 0, n \geq 0$$

$$V_n(x,t)dx = P\{C(t) = 3, X(t) = n, x \leq \xi_3(t) \leq x + dx\}, t \geq 0, x \geq 0, n \geq 0$$

$$R_n(x,t)dx = P\{C(t) = 4, X(t) = n, x \leq \xi_4(t) \leq x + dx\}, t \geq 0, x \geq 0, n \geq 0$$

where, $P_n(x,t)$ = Probability that at time t, there are n customers in the orbit and the elapsed retrial time is x. $Q_n^i(x,t)$ = Probability that at time t, there are n customers in the orbit excluding one customer in the ith type of service and the elapsed service time for this customer is x. Consequently $Q_n^i(t) = \int_0^\infty Q_n^i(x,t)$ denotes the probability that at time t there are n customers in the queue excluding the one who is getting i^{th} type of service irrespective of the value of x. $V_n(x,t)$ = Probability that there are n customers in the orbit and the elapsed vacation time is x. $R_n(t)$ = Probability that there are n customers in the orbit and the elapsed repair time is x. $P_0(t)$ = Probability that there are zero customers in the orbit and the server is free.

Based on the above assumptions and notations, our model is governed by the following set of differential difference equations,

$$\frac{dP_0(t)}{dt} = -\lambda P_0(t) + \int_0^\infty V_0(x,t)\eta(x)dx \tag{1}$$

$$\frac{\partial P_n(x,t)}{\partial t} + \frac{\partial P_n(x,t)}{\partial x} = -(\lambda + \theta(x))P_n(x,t) \tag{2}$$

$$\frac{\partial Q_0^1(x,t)}{\partial t} + \frac{\partial Q_0^1(x,t)}{\partial x} = -(\lambda + \mu_1(x))Q_0^1(x,t) + \lambda(1-b)Q_0^1(x,t) \tag{3}$$

$$\frac{\partial Q_n^1(x,t)}{\partial t} + \frac{\partial Q_n^1(x,t)}{\partial x} = -(\lambda + \mu_1(x))Q_n^1(x,t) + \lambda b Q_{n-1}^1(x,t) + \lambda(1-b)Q_n^1(x,t) \quad (4)$$

$$\frac{\partial Q_0^2(x,t)}{\partial t} + \frac{\partial Q_0^2(x,t)}{\partial x} = -(\lambda + \mu_2(x))Q_0^2(x,t) + \lambda(1-b)Q_0^2(x,t) \quad (5)$$

$$\frac{\partial Q_n^2(x,t)}{\partial t} + \frac{\partial Q_n^2(x,t)}{\partial x} = -(\lambda + \mu_2(x))Q_n^2(x,t) + \lambda b Q_{n-1}^2(x,t) + \lambda(1-b)Q_n^2(x,t) \quad (6)$$

$$\frac{\partial V_0(x,t)}{\partial t} + \frac{\partial V_0(x,t)}{\partial x} = -(\lambda + \eta(x))V_0(x,t) + \lambda(1-b)V_0(x,t) \quad (7)$$

$$\frac{\partial V_n(x,t)}{\partial t} + \frac{\partial V_n(x,t)}{\partial x} = -(\lambda + \eta(x))V_n(x,t) + \lambda b V_{n-1}(x,t) + \lambda(1-b)V_n(x,t) \quad (8)$$

$$\frac{\partial R_0(x,t)}{\partial t} + \frac{\partial R_0(x,t)}{\partial x} = -(\lambda + v(x))R_0(x,t) + \lambda(1-b)R_0(x,t) \quad (9)$$

$$\frac{\partial R_n(x,t)}{\partial t} + \frac{\partial R_n(x,t)}{\partial x} = -(\lambda + v(x))R_n(x,t) + \lambda b R_{n-1}(x,t) + \lambda(1-b)R_n(x,t) \quad (10)$$

The above equations have been solved subject to the following boundary equations:

$$P_n(0,t) = p \int_0^\infty Q_n^1(x,t)\mu_1(x)dx + p \int_0^\infty Q_n^2(x,t)\mu_2(x)dx + \int_0^\infty V_n(x,t)\eta(x)dx$$

$$+ \int_0^\infty R_n(x,t)v(x)dx \quad (11)$$

$$Q_0^1(0,t) = \alpha\theta_1 \int_0^\infty P_1(x,t)\theta(x)dx + \alpha\theta_1 \lambda P_0(t) \quad (12)$$

$$Q_n^1(0,t) = \alpha\theta_1 \int_0^\infty P_{n+1}(x,t)\theta(x)dx + \alpha\theta_1 \lambda \int_0^\infty P_n(x,t)dx \quad (13)$$

$$Q_0^2(0,t) = \alpha\theta_2 \int_0^\infty P_1(x,t)\theta(x)dx + \alpha\theta_2 \lambda P_0(t) \quad (14)$$

$$Q_n^2(0,t) = \alpha\theta_2 \int_0^\infty P_{n+1}(x,t)\theta(x)dx + \alpha\theta_2 \lambda \int_0^\infty P_n(x,t)dx \quad (15)$$

$$V_0(0,t) = \int_0^\infty Q_0^1(x,t)\mu_1(x)dx + \int_0^\infty Q_0^2(x,t)\mu_2(x)dx \quad (16)$$

$$V_n(0,t) = q \int_0^\infty Q_n^1(x,t)\mu_1(x)dx + q \int_0^\infty Q_n^2(x,t)\mu_2(x)dx \quad (17)$$

$$R_1(0,t) = \bar{\alpha} \int_0^\infty P_1(x,t)\theta(x)dx + \bar{\alpha}\lambda P_0(t) \tag{18}$$

$$R_n(0,t) = \bar{\alpha} \int_0^\infty P_n(x,t)\theta(x)dx + \bar{\alpha}\lambda \int_0^\infty P_{n-1}(x,t)dx \tag{19}$$

5 The Steady State Solution

The system of differential difference equations to describe the $M/G/1$ retrial queue with balking, two types of service, Bernoulli vacation and starting failure are given in Eqs. (1)–(19) along with the stability condition $(\alpha\lambda b\theta_1\beta_1^1 + \alpha\lambda b\theta_2\beta_1^2 + \alpha\lambda bqV_1 + \bar{\alpha}(1 + \lambda bR_1)) < \bar{A}(\lambda)$, the stationary distributions of the number of jobs in the system when the server being idle, busy with type 1 service, busy with type 2 service and on vacations are

$$Dr = [\alpha(p + q\bar{V}(\lambda b(1 - z)))(\theta_1\bar{B}_1(\lambda b(1 - z)) + \theta_2\bar{B}_2(\lambda b(1 - z))) + z\bar{\alpha}\bar{R}(\lambda b(1 - z))]$$
$$[z + (1 - z)\bar{A}(\lambda)] - z$$

$$P(z) = \frac{P_0 z(1 - \bar{A}(\lambda))}{\bar{V}(\lambda b)}$$
$$\frac{\bar{V}(\lambda b)[1 - \{\alpha(p + q\bar{V}(\lambda b(1 - z)))(\theta_1\bar{B}_1(\lambda b(1 - z)) + \theta_2\bar{B}_2(\lambda b(1 - z))) + z\bar{\alpha}\bar{R}(\lambda b(1 - z))\}]}{+ p(1 - \bar{V}(\lambda b(1 - z)))}$$
$$\overline{ Dr }$$

$$Q^1(z) = \frac{\alpha\theta_1 P_0[1 - \bar{B}_1(\lambda b(1 - z))]}{\bar{V}(\lambda b)b(1 - z)}$$
$$\left\{ \frac{p(1 - \bar{V}(\lambda b(1 - z)))(z + (1 - z)\bar{A}(\lambda)) + (1 - z)\bar{A}(\lambda)\bar{V}(\lambda b)}{Dr} \right\} \tag{21}$$

$$Q^2(z) = \frac{\alpha\theta_2 P_0[1 - \bar{B}_2(\lambda b(1 - z))]}{\bar{V}(\lambda b)b(1 - z)}$$
$$\left\{ \frac{p(1 - \bar{V}(\lambda b(1 - z)))(z + (1 - z)\bar{A}(\lambda)) + (1 - z)\bar{A}(\lambda)\bar{V}(\lambda b)}{Dr} \right\} \tag{22}$$

$$V(z) = \frac{P_0(1 - \bar{V}(\lambda b(1 - z)))}{b(1 - z)\bar{V}(\lambda b)} \frac{\left\{ \theta_1\bar{B}_1(\lambda b(1 - z)) + \theta_2\bar{B}_2(\lambda b(1 - z)) \right\} \left\{ \begin{matrix} \alpha p(z + (1 - z)\bar{A}(\lambda)) + \\ \alpha q(1 - z)\bar{V}(\lambda b)\bar{A}(\lambda) \end{matrix} \right\}}{+ pz\bar{\alpha}\bar{R}(\lambda b(1 - z))(z + (1 - z)\bar{A}(\lambda)) - pz}}{Dr} \tag{23}$$

$$R(z) = \frac{z\bar{\alpha}P_0[1 - \bar{R}(\lambda b(1 - z))]}{\bar{V}(\lambda b)b(1 - z)}$$
$$\left\{ \frac{p(1 - \bar{V}(\lambda b(1 - z)))(z + (1 - z)\bar{A}(\lambda)) + (1 - z)\bar{A}(\lambda)\bar{V}(\lambda b)}{Dr} \right\} \tag{24}$$

From the above equations, the only unknown is P_0 which can be determined by using the normalization condition $P(1) + Q^1(1) + Q^2(1) + V(1) + R(1) + P_0 = 1$. Thus by substituting $z = 1$ in the above equations and applying L'Hopital's rule wherever necessary, also using some algebraic manipulations we get

$$P_0 = \frac{b\bar{V}(\lambda b)\left[\bar{A}(\lambda) - (\alpha\lambda bqV_1 + \alpha\lambda b\theta_1\beta_1^1 + \alpha\lambda b\theta_2\beta_2^1) - \bar{\alpha}(1 + \lambda bR_1)\right]}{\left\{\begin{array}{l} (1-b)\left[p\lambda bV_1\bar{A}(\lambda) + \bar{A}(\lambda)\bar{V}(\lambda b)\left[\alpha\lambda b\theta_1\beta_1^1 + \alpha\lambda b\theta_2\beta_2^1 + \alpha\lambda bqV_1 + \bar{\alpha}\lambda bR_1\right]\right] \\ + \alpha b\bar{A}(\lambda)\bar{V}(\lambda b) + (b-\bar{\alpha})p\lambda bV_1 \end{array}\right\}}$$

(25)

The probability generating function of the number of customers in the system $S(z) = P(z) + zQ^1(z) + zQ^2(z) + V(z) + R(z) + P_0$ is obtained from the above equations as,

$$S(z) = \frac{\left\{\begin{array}{l} \left\{\alpha p(z + (1-z)\bar{A}(\lambda))(1 - \bar{V}(\lambda b(1-z))) + \alpha(1-z)\bar{V}(\lambda b)\bar{A}(\lambda)\right\}\left\{\begin{array}{l}\theta_1\overline{B_1}(\lambda b(1-z)) \\ + \theta_2\overline{B_2}(\lambda b(1-z))\end{array}\right\} \\ + (1-b)\left\{\begin{array}{l} z\bar{A}(\lambda)\bar{V}(\lambda b) - pz(1 - \bar{A}(\lambda))(1 - \bar{V}(\lambda b(1-z))) - \bar{\alpha}z\bar{A}(\lambda)\bar{V}(\lambda b)\bar{R}(\lambda b(1-z)) \\ -\alpha(p + q\bar{V}(\lambda b(1-z)))\bar{A}(\lambda)\bar{V}(\lambda b)\{\theta_1\overline{B_1}(\lambda b(1-z)) + \theta_2\overline{B_2}(\lambda b(1-z))\} \end{array}\right\} \end{array}\right\}}{b\bar{V}(\lambda b)\{Dr\}}$$

(26)

Define $O(z) = P(z) + Q^1(z) + Q^2(z) + V(z) + R(z) + P_0$. Thus $O(z)$ represents the probability generating function for the number of customers in the orbit. Using the above equations and simplifying them, we obtain

$$O(z) = \frac{\left\{\begin{array}{l} \left\{\alpha p(z + (1-z)\bar{A}(\lambda))(1 - \bar{V}(\lambda b(1-z))) + \alpha(1-z)\bar{V}(\lambda b)\bar{A}(\lambda)\right\} \\ + (1-b)\left\{\begin{array}{l} z\bar{A}(\lambda)\bar{V}(\lambda b) - pz(1 - \bar{A}(\lambda))(1 - \bar{V}(\lambda b(1-z))) - \bar{\alpha}z\bar{A}(\lambda)\bar{V}(\lambda b)\bar{R}(\lambda b(1-z)) \\ -\alpha(p + q\bar{V}(\lambda b(1-z)))\bar{A}(\lambda)\bar{V}(\lambda b)\{\theta_1\overline{B_1}(\lambda b(1-z)) + \theta_2\overline{B_2}(\lambda b(1-z))\} \end{array}\right\} \end{array}\right\}}{b\bar{V}(\lambda b)\{Dr\}}$$ (27)

6 Performance Measures

In this section we derive some of the performance measures of the $M/G/1$ retrial queue with two types of service, balking, Bernoulli vacation and starting failure. Let E_1, E_2, I, C and E denotes the server utilization of type 1 service, the server utilization of type 2 service, the steady state probability that the server is idle during the retrial time, the steady state probability that the server is on vacation and the steady state probability that the server is on repair.

$$I = P(1)$$
$$= \frac{P_0(1 - \bar{A}(\lambda))\left[p\lambda bV_1 + \bar{V}(\lambda b)\left(\alpha q\lambda bV_1 + \alpha\theta_1\lambda b\beta_1^1 + \alpha\theta_2\lambda b\beta_2^1 + \bar{\alpha} + \bar{\alpha}\lambda bR_1\right)\right]}{\bar{V}(\lambda b)\left[\bar{A}(\lambda) - (\alpha q\lambda bV_1 + \alpha\theta_1\lambda b\beta_1^1 + \alpha\theta_2\lambda b\beta_2^1 + \bar{\alpha} + \bar{\alpha}\lambda bR_1)\right]}$$

$$E_1 = Q^1(1) = \frac{\alpha\theta_1 P_0\left[\lambda b\beta_1^1(\lambda bpV_1 + \bar{A}(\lambda)\bar{V}(\lambda b))\right]}{b\bar{V}(\lambda b)\left[\bar{A}(\lambda) - (\alpha q\lambda bV_1 + \alpha\theta_1\lambda b\beta_1^1 + \alpha\theta_2\lambda b\beta_2^1 + \bar{\alpha} + \bar{\alpha}\lambda bR_1)\right]}$$

$$E_2 = Q^2(1) = \frac{\alpha\theta_2 P_0\left[\lambda b\beta_2^1(\lambda bpV_1 + \bar{A}(\lambda)\bar{V}(\lambda b))\right]}{b\bar{V}(\lambda b)\left[\bar{A}(\lambda) - (\alpha q\lambda bV_1 + \alpha\theta_1\lambda b\beta_1^1 + \alpha\theta_2\lambda b\beta_2^1 + \bar{\alpha} + \bar{\alpha}\lambda bR_1)\right]}$$

$$C = V(1) = \frac{\lambda bV_1 P_0\left[p\bar{A}(\lambda) + \alpha q\bar{V}(\lambda b)\bar{A}(\lambda) - \alpha p(\theta_1\lambda b\beta_1^1 + \theta_2\lambda b\beta_2^1) - \bar{\alpha}p - \bar{\alpha}p\lambda bR_1\right]}{b\bar{V}(\lambda b)\left[\bar{A}(\lambda) - (\alpha q\lambda bV_1 + \alpha\theta_1\lambda b\beta_1^1 + \alpha\theta_2\lambda b\beta_2^1 + \bar{\alpha} + \bar{\alpha}\lambda bR_1)\right]}$$

$$E = R(1) = \frac{\bar{\alpha}P_0\lambda bR_1(p\lambda bV_1 + \bar{A}(\lambda)\bar{V}(\lambda b))}{b\bar{V}(\lambda b)\left[\bar{A}(\lambda) - (\alpha q\lambda bV_1 + \alpha\theta_1\lambda b\beta_1^1 + \alpha\theta_2\lambda b\beta_2^1 + \bar{\alpha} + \bar{\alpha}\lambda bR_1)\right]}$$

Let L_q denote the mean number of customers in the orbit under the steady state,

$$L_q = \frac{d}{dz}O(z)\Big|_{z=1}$$

$$L_q = \frac{\left\{\begin{array}{l} 2\alpha p\lambda bV_1(1 - \bar{A}(\lambda)) + \alpha p\lambda^2 b^2 V_2 - (1-b)\{2p\lambda bV_1(1 - \bar{A}(\lambda)) + p\lambda^2 b^2 V_2(1 - \bar{A}(\lambda))\} \\ + (1-b)\left[\bar{A}(\lambda)\bar{V}(\lambda b)\left\{\begin{array}{l}2\alpha q\lambda bV_1(\theta_1\lambda b\beta_1^1 + \theta_2\lambda b\beta_2^1) + \alpha(\theta_1\lambda^2 b^2\beta_1^2 + \theta_2\lambda^2 b^2\beta_2^2)\\ + q\alpha\lambda^2 b^2 V_2 + 2\bar{\alpha}\lambda bR_1 + \bar{\alpha}\lambda^2 b^2 R_2\end{array}\right\}\right]\end{array}\right\}}{2\left\{\alpha b\bar{A}(\lambda)\bar{V}(\lambda b) + (b-\bar{\alpha})p\lambda bV_1 + (1-b)\left[\begin{array}{l}p\lambda bV_1\bar{A}(\lambda) + \bar{A}(\lambda)\bar{V}(\lambda b)\\ \left[\alpha\lambda b\theta_1\beta_1^1 + \alpha\lambda b\theta_2\beta_2^1 + \alpha\lambda bqV_1 + \bar{\alpha}\lambda bR_1\right]\end{array}\right]\right\}} +$$

$$\frac{\left\{b\bar{V}(\lambda b)\left\{\begin{array}{l}q\alpha\lambda^2 b^2 V_2 + 2\alpha q\lambda bV_1(\theta_1\lambda b\beta_1^1 + \theta_2\lambda b\beta_2^1) + \alpha(\theta_1\lambda^2 b^2\beta_1^2 + \theta_2\lambda^2 b^2\beta_2^2) + 2\bar{\alpha}\lambda bR_1\\ + \bar{\alpha}\lambda^2 b^2 R_2 + 2\left[\alpha\lambda bqV_1 + \alpha\lambda b\theta_1\beta_1^1 + \alpha\lambda b\theta_2\beta_2^1 + \bar{\alpha} + \bar{\alpha}\lambda bR_1\right]\left[1 - \bar{A}(\lambda)\right]\end{array}\right\}\right\}}{2b\bar{V}(\lambda b)\left[\bar{A}(\lambda) - \left[\alpha\lambda bqV_1 + \alpha\lambda b\theta_1\beta_1^1 + \alpha\lambda b\theta_2\beta_2^1 + \bar{\alpha} + \bar{\alpha}\lambda bR_1\right]\right]}$$

Similarly L_s denote the mean number of customers in the system under the steady state,

$$L_s = \frac{d}{dz}S(z)\Big|_{z=1}$$

$$L_s = \frac{\left\{\begin{array}{l} 2\alpha p\lambda bV_1(1 - \bar{A}(\lambda)) + \alpha p\lambda^2 b^2 V_2 - (1-b)\{2p\lambda bV_1(1 - \bar{A}(\lambda)) + p\lambda^2 b^2 V_2(1 - \bar{A}(\lambda))\} \\ + 2[\alpha p\lambda bV_1 + \alpha\bar{V}(\lambda b)\bar{A}(\lambda)]\left[\theta_1\lambda b\beta_1^1 + \theta_2\lambda b\beta_2^1\right] \\ + (1-b)\left[\bar{A}(\lambda)\bar{V}(\lambda b)\left\{\begin{array}{l}2\alpha q\lambda bV_1(\theta_1\lambda b\beta_1^1 + \theta_2\lambda b\beta_2^1) + \alpha(\theta_1\lambda^2 b^2\beta_1^2 + \theta_2\lambda^2 b^2\beta_2^2)\\ + q\alpha\lambda^2 b^2 V_2 + 2\bar{\alpha}\lambda bR_1 + \bar{\alpha}\lambda^2 b^2 R_2\end{array}\right\}\right]\end{array}\right\}}{2\left\{\alpha b\bar{A}(\lambda)\bar{V}(\lambda b) + (b-\bar{\alpha})p\lambda bV_1 + (1-b)\left[\begin{array}{l}p\lambda bV_1\bar{A}(\lambda) + \bar{A}(\lambda)\bar{V}(\lambda b)\\ \left[\alpha\lambda b\theta_1\beta_1^1 + \alpha\lambda b\theta_2\beta_2^1 + \alpha\lambda bqV_1 + \bar{\alpha}\lambda bR_1\right]\end{array}\right]\right\}} +$$

$$\frac{\left\{b\bar{V}(\lambda b)\left\{\begin{array}{l}q\alpha\lambda^2 b^2 V_2 + 2\alpha q\lambda bV_1(\theta_1\lambda b\beta_1^1 + \theta_2\lambda b\beta_2^1) + \alpha(\theta_1\lambda^2 b^2\beta_1^2 + \theta_2\lambda^2 b^2\beta_2^2) + 2\bar{\alpha}\lambda bR_1\\ + \bar{\alpha}\lambda^2 b^2 R_2 + 2\left[\alpha\lambda bqV_1 + \alpha\lambda b\theta_1\beta_1^1 + \alpha\lambda b\theta_2\beta_2^1 + \bar{\alpha} + \bar{\alpha}\lambda bR_1\right]\left[1 - \bar{A}(\lambda)\right]\end{array}\right\}\right\}}{2b\bar{V}(\lambda b)\left[\bar{A}(\lambda) - \left[\alpha\lambda bqV_1 + \alpha\lambda b\theta_1\beta_1^1 + \alpha\lambda b\theta_2\beta_2^1 + \bar{\alpha} + \bar{\alpha}\lambda bR_1\right]\right]}$$

7 Special Cases

In this section we present some of the special cases for the proposed model.

Case 1: If $b = 1$, $\alpha = 1$ then our model can be reduced to $M/G/1$ retrial queue with two types of service, Bernoulli vacation and general retrial times by Arivudainambi and Gowsalya (in press) [3]

Case 2: If $\theta_2 = 0$, $b = 1$, $\alpha = 1$ then our model can be reduced to $M/G/1$ retrial queue with Bernoulli schedules and general retrial times by Krishna Kumar and Arivudainambi [12].

Case 3: If $\bar{A}(\lambda) \to 1$, $\theta_2 = 0$, $b = 1$, $\alpha = 1$ and $p = 1$, then the system reduces to $M/G/1$ queue with single vacation model by Doshi [6]. Thus the probability generating function of the number of customers in the system $S(z)$ becomes

Case 4: If $\bar{A}(\lambda) \to 1$, $\theta_2 = 0$, $b = 1$, $\alpha = 1$ and $q = 1$, then the system reduces to $M/G/1$ queue one limited service system with single vacation policy by Takagi [17]. Thus the probability generating function of the number of customers in the system $S(z)$ becomes

Case 5: If $\bar{A}(\lambda) \to 1$, $\theta_2 = 0$, $b = 1$, $\alpha = 1$ and $q = 0$, then the system reduces to the classical $M/G/1$ queueing system, i.e., the well-known Pollaczek-Khintchine formula.

8 Cost Analysis

This section deals with the optimum design of the proposed model by using the usual cost notation structure and the expected total cost per unit of time is given by,

$$ETC = C_h L_s + C_{b_1} E_1 + C_{b_2} E_2 + C_r E + C_v C + C_I I$$

where

- C_h = Holding cost per unit customer
- C_{b_1} = Cost per unit time while serving type 1 service for the customers
- C_{b_2} = Cost per unit time while serving type 2 service for the customers
- C_r = Cost per unit time for providing repair to the failed server
- C_v = Cost per unit time in the system when the server is on vacation
- C_I = Cost per unit time when the customer retry for the service

The total cost function nature on the various parameters have been studied so as to visualize the ETC nature. Arbitrarily choosing the cost element default values as $C_h = \$5$, $C_{b_1} = \$1500$, $C_{b_2} = \$1000$, $C_r = \$200$, $C_v = \$100$, and $C_I = \$200$ and other parameters $\lambda = 1$, $\theta_1 = 0.5$, $\theta_2 = 0.5$, $\theta = 11$, $\mu_1 = 45$, $\mu_2 = 30$, $v = 40$, $\eta = 35$, $b = 0.2$, $\alpha = 0.2$, $\bar{\alpha} = 0.8$, $p = 0.1$ and $q = 0.9$. Tables 1, 2, 3, 4 and 5 shows the effects of (C_{b_1}, C_h), (C_h, C_{b_2}), (C_v, C_r), (C_r, C_I) and (C_v, C_I) on the expected cost function and illustrates a linearly increasing trend against the increasing cost parameters. From the examined tables we find the optimum expected total cost is \$131.0228.

Table 1. Effects of (C_h, C_{b_1}) on the expected total cost function ETC with $C_{b_2} = \$1000$ $C_v = \$100$, $C_r = \$200$, $C_i = \$200$

(C_h, C_{b_1})	(5, 1500)	(5, 1550)	(5, 1600)	(10, 1500)	(15, 1500)
ETC	131.0228	131.5170	132.0111	146.9664	162.9099

Table 2. Effects of (C_h, C_{b_2}) on the expected total cost function ETC with $C_{b_1} = \$1500$, $C_v = \$100$, $C_r = \$200$, $C_i = \$200$

(C_h, C_{b_2})	(5, 1000)	(5, 1050)	(5, 1100)	(10, 1000)	(15, 1000)
ETC	131.0228	131.7640	132.5053	146.9664	162.9099

Table 3. Effects of (C_v, C_r) on the expected total cost function ETC with $C_h = \$5$, $C_{b_1} = \$1500$, $C_{b_2} = \$1000$, $C_i = \$200$

(C_v, C_r)	(100, 200)	(150, 200)	(200, 200)	(100, 250)	(100, 300)
ETC	131.0228	132.2435	133.4642	135.4705	139.9181

Table 4. Effects of (C_r, C_i) on the expected total cost function ETC with $C_h = \$5$, $C_{b_1} = \$1500$, $C_{b_2} = \$1000$, $C_v = \$100$

(C_r, C_i)	(200, 200)	(250, 200)	(300, 200)	(200, 250)	(200, 300)
ETC	131.0228	135.4705	139.9181	147.3217	163.6207

Table 5. Effects of (C_v, C_i) on the expected total cost function ETC with $C_h = \$5$, $C_{b_1} = \$1500$, $C_{b_2} = \$1000$, $C_r = \$200$

(C_v, C_i)	(100, 200)	(150, 200)	(200, 200)	(100, 250)	(100, 300)
ETC	131.0228	132.2435	133.4642	147.3217	163.6207

9 Conclusion and Future Work

The authors have studied retrial queueing system with two types of service, Balking, Bernoulli schedules and starting failure. Cost analysis for the proposed model have been investigated. The primary or repeated customers who arrives for service have an option to choose either type 1 service with probability θ_1 or type 2 service with probability θ_2 under Bernoulli vacation policy. The necessary and sufficient condition for the proposed model have been derived. Sensitivity analysis have been carried out by various performance measures of the characteristics of the server, numerical illustrations have been provided for different parameters on the obtained result. The authors have provided explicit expressions like expected number of customers in the system and expected number of customers in the orbit have been obtained by the use of supplementary variable technique. The proposed model have also been examined through different vacation policies like, single vacation, modified vacation policy, working vacation and different types of service like optional service, two phases of service and considering feedback models into account which we hope to examine in future.

References

1. Artalejo, J.R.: Accessible bibliography on retrial queues: progress in 2000–2009. Math. Comput. Model. **51**, 1071–1081 (2010)
2. Arivudainambi, D., Godhandaraman, P.: Retrial queueing system with balking, optional service and vacation. Ann. Oper. Res. **229**, 67–84 (2015)
3. Arivudainambi, D., Gowsalya, M.: A single server non-markovian retrial queue with two types of service and Bernoulli vacation. Int. J. Oper. Res. (in Press)
4. Artalejo, J.R., Gomez-Corral, A.: Steady state solution of a single server queue with linear repeated request. J. Appl. Probab. **1**, 223–233 (1997)
5. Choudhury, G., Ke, J.: An unreliable retrial queue with delaying repair and general retrial times under Bernoulli vacation schedule. Appl. Math. Comput. **230**, 436–450 (2014)
6. Doshi, B.T.: Queueing systems with vacations-a survey. Queueing Syst. **1**, 29–66 (1986)
7. Fayolle, G.: A simple telephone exchange with delayed feedback. In: Boxma, O.J., Cohen, J. W., Tijms, H.C. (eds.) Teletraffic Analysis and Computer Performance Evaluation Elsevier Amsterdam, vol. 7, pp. 245–253 (1986)
8. Kapyrin, V.A.: A study of the stationary distributions of a queueing system with recurring demands. Cybernetics **13**, 584–590 (1997)
9. Ke, J., Wu, C.H., Zhang, Z.G.: Recent developments in vacation queueing models: a short survey. Int. J. Oper. Res. **7**, 3–8 (2010)
10. Ke, J., Chang, F.: Modified vacation policy for $M/G/1$ retrial queue with balking and feedback. Comput. Ind. Eng. **57**, 433–443 (2009)
11. Keilson, J., Servi, L.D.: Oscillating random walk models for $G/G/1$ vacation systems with Bernoulli schedules. J. Appl. Probab. **3**, 790–802 (1986)
12. Krishna Kumar, B., Arivudainambi, D.: The $M/G/1$ retrial queue with Bernoulli schedules and general retrial times. Comput. Math Appl. **43**, 15–30 (2002)
13. Krishna Kumar, B., Pavai Madheswari, S., Vijayakumar, A.: The M/G/1 retrial queue with feedback and starting failures. Appl. Math. Model. **26**, 1057–1075 (2002)
14. Rajadurai, P., Saravanarajan, M.C., Chandrasekaran, V.M.: Analysis of repairable $M^{[X]}/(G_1,G_2)/1$- feedback retrial G-queue with balking and starting failures under at most J vacations. Appl. Appl. Mathe. **10**, 694–717 (2015)
15. Sennot, L.I., Humblet, P.A., Tweedie, R.L.: Mean drift and the non ergodicity of Markov chains. Oper. Res. **31**, 783–789 (1983)
16. Sumitha, D., Udaya Chandrika, K.: Retrial queuing system with starting failure, single vacation and orbital search. Int. J. Comput. Appl. **40**, 29–33 (2012)
17. Takagi, H.: Queueing Analysis. Vacation and Priority Systems, vol. 1, North-Holland Amsterdam (1991)
18. Yang, T., Templeton, J.G.C.: A survey on retrial queues. Queueing Syst. **2**, 201–233 (1987)

Mathematical Modelling and Optimization of Lung Cancer in Smokers and Non Smokers

Elizabeth Sebastian$^{(\boxtimes)}$ and Priyanka Victor

Auxilium College (Autonomous), Vellore, Tamil Nadu, India
elizafma@gmail.com, priyankavictor2@gmail.com

Abstract. Cancer epidemiology is the division of epidemiology apprehensive with the disease cancer. It makes use of epidemiological methods to discover the reason for cancer and to recognize and discover advanced treatments. In this paper, we construct and analyze a discrete time mathematical model on lung cancer involving smokers and non smokers. We derive the two equilibrium points namely smoke free and smoke induced equilibrium and analyze the conditions in which the equilibrium points are stable or unstable. We also find the optimality conditions using Pontryagin's Principle. Finally, we prove our theoretical results using numerical simulations through MATLAB.

Keywords: Difference equations · Smoke-free equilibrium
Smoke induced equilibrium · Optimization · Lung cancer

1 Introduction

A mathematical model is a description of a system using mathematical concepts. Epidemic models are used as a tool to analyze the behaviors of biological diseases and how they spread. Lung cancer is the leading cause of cancer deaths worldwide. The WHO reports that over 1.1 million people die of Lung cancer each year. As a result, WHO has identified Lung cancer as one of the new problems facing the world in this new century.

Majority of lung cancer occur in people who are either current or former smokers. While the relationship between smoking and lung cancer is well established, other factors also came into play. The health risks of tobacco smoke are not limited to smokers. The lungs of anyone who breathes the air that contains tobacco smoke are exposed to its carcinogens. Therefore, exposure to smoky air in the home, workplace, or in public can increase a person's risk of lung cancer. This kind of exposure is called second-hand smoke, side-stream smoke, environmental tobacco smoke or passive smoke. Some of the most common lung carcinogens are asbestos, radon, arsenic, chromium and nickel. We consider the passive smoke as an epidemic as it spreads in a way similar to infections and anyone near it can be affected.

Global Dynamics of a Mathematical model on smoking with Media Campaigns was studied by Verma and Agarwal [1]. A Mathematical model for Lung Cancer: The effects of Second Hand smoke and Education was studied by Acevedo-Estefania et al. [2]. A Local Stability of Mathematical Models for Cancer Treatment by using Gene

G. Ganapathi et al. (Eds.): ICC3 2017, CCIS 844, pp. 205–213, 2018.
https://doi.org/10.1007/978-981-13-0716-4_17

Therapy was studied by Lestari and Ambarwati [4]. In this paper, we construct a mathematical model on Lung cancer involving smokers and non smokers by modelling the effect of passive smoke in non smokers and optimise the cost of awareness campaigns. In Sect. 2, we formulate the model. Equilibrium points of the model are derived in Sect. 3. Stability analysis is done in Sect. 4. In Sect. 5, we provide the optimal control problem of the model. Numerical example is given in Sect. 6.

2 Model Formulation

We construct a mathematical model using a system of Differential equations which is given below [1–3] (Fig. 1):

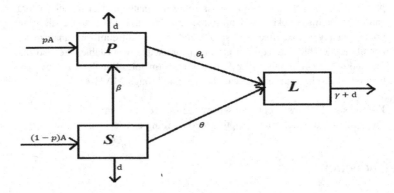

Fig. 1. Compartmental model of Lung cancer epidemic

$$P_{t+1} = P_t + pA + \beta P_t S_t - (\theta_1 + d)P_t$$
$$S_{t+1} = S_t + (1 - p)A - \beta P_t S_t - (\theta + d)S_t \qquad (1)$$
$$L_{t+1} = L_t + \theta S_t + \theta_1 P_t - (\gamma + d)L_t$$

where
P_t is the population of non smokers at time t. S_t is the population of smokers at time t. L_t is the population affected by lung cancer at time t. A is the constant population. p is the rate of population who do not smoke. β is the rate of passive smoke intake by non smokers when they come in contact with smokers. d is the natural death rate. γ is the death rate due to lung cancer. δ_1 is the death rate due to lung cancer with treatment. θ is the incidence rate of smokers. θ_1 is the incidence rate of non smokers.
We make the following assumptions:

- Our population consists of smokers and non-smokers. Smokers are in contact with non-smokers.
- The non-smokers get lung cancer due to passive smoking.
- The rate at which non smokers get lung cancer is directly proportional to the passive smoke they are exposed to.

Let us take $N_t = P_t + S_t + L_t$, Adding all the equations of the model, we get [4, 5]

$$N(t+1) = N(t) + A - dN(t+1) - \delta L(t) - \delta_1 T(t)$$
$$\leq N(t) + A - dN(t+1) \tag{2}$$

For our model, we get the equilibrium point $N^* = \frac{A}{d}$, which is globally asymptotically stable as $\lim_{t \to \infty} N(t) = N^*$.

The initial conditions are given by,

$$P(0) \geq 0, S(0) \geq 0, L(0) \geq 0 \tag{3}$$

Let us assume that the following condition holds:

$$\beta P^0 + \theta + d, \ \gamma + d, \ \theta_1 + d \geq 0 \tag{4}$$

3 Equilibrium Points

The system has two equilibrium points namely the smoke free equilibrium and the smoke induced equilibrium [4].

- **Smoke- free Equilibrium**: Smoke free equilibrium is the condition in which there is no passive intake of smoke. $E^0 = (P^0, 0, 0)$
  where $P^0 = \frac{pA}{\theta_1 + d}$
- **Smoke- Induced Equilibrium**: Smoke Induced equilibrium is the condition in which there is passive intake of smoke. $E^* = (P^*, S^*, L^*)$

  where

  $$P^* = \frac{(\theta + d)pA}{\beta A + (1 + \theta + d)(\theta_1 + d)}$$
  $$S^* = \frac{(1 - p)A}{\beta pA(\theta + d) + [\beta A + (1 + \theta + d)(\theta_1 + d)](\theta + d)}$$
  $$L^* = \frac{\theta S^* + \theta_1 P^*}{(\gamma + d)}$$

4 Stability Analysis

Theorem 1: The smoke free equilibrium is locally asymptotically stable if condition (4) holds.

Proof: The Jacobian matrix of System (1) at E^0 is given by

$$J(E^0) = \begin{pmatrix} 1-(\theta_1+d) & \beta P^0 & 0 \\ 0 & 1-\beta P^0 - (\theta+d) & 0 \\ \theta_1 & \theta & 1-(\gamma+d) \end{pmatrix} \tag{5}$$

The eigen values of this matrix is given by

$$\lambda_1 = 1-(\theta_1+d), \lambda_2 = 1-\beta P^0 - (\theta+d), \lambda_3 = 1-(\gamma+d)$$

The modulus of the eigen values is less than one if the conditions (4) are satisfied. Hence the smoke free equilibrium is locally asymptotically stable [6–8].

Theorem 2: The smoke induced equilibrium is locally asymptotically stable if

$$[1+\beta S^* - (\theta_1+d)] + [1-\beta P^* - (\theta+d)] > 0$$
$$[1+\beta S^* - (\theta_1+d)][1-\beta P^* - (\theta+d)] + \beta^2 P^* S^* < 0$$

Otherwise unstable.

Proof: The Jacobian matrix of system (1) at E^* is given by

$$J(E^*) = \begin{pmatrix} 1+\beta S^* - (\theta_1+d) & \beta P^* & 0 \\ -\beta S^* & 1-(\theta+d)-\beta P^* & 0 \\ \theta_1 & \theta & 1-(\gamma+d) \end{pmatrix} \tag{6}$$

One of the eigen values of the matrix is given by

$$\lambda = 1-(\gamma+\delta+d)$$

The remaining matrix can be written as,

$$J(E^*) = \begin{pmatrix} 1+\beta S^* - (\theta_1+d) & \beta P^* \\ -\beta S^* & 1-(\theta+d)-\beta P^* \end{pmatrix} \tag{7}$$

The characteristic equation of the matrix is given by,

$$\varphi(\lambda) = \lambda^2 + a_1\lambda + a_2 = 0 \tag{8}$$

where

$$a_1 = -[[1+\beta S^* - (\theta_1+d)] + [1-\beta P^* - (\theta+d)]]$$
$$a_2 = [1+\beta S^* - (\theta_1+d)][1-\beta P^* - (\theta+d)] + \beta^2 P^* S^*$$

We see that characteristic Eq. (8) has positive roots if

$$[1 + \beta S^* - (\theta_1 + d)] + [1 - \beta P^* - (\theta + d)] > 0$$
$$[1 + \beta S^* - (\theta_1 + d)][1 - \beta P^* - (\theta + d)] + \beta^2 P^* S^* < 0$$

5 Optimization

Our aim is to reduce the number of lung cancer incidence through public awareness during the time steps $t = 0\,to\,T$ and also minimizing the cost spent in the public awareness programs. We are assuming the cost of administering the control is quadratic for simplicity. The problem is to minimize the objective functional [2, 3, 5]:

$$J(u) = A_T S_T + \sum_{k=0}^{T-1} \left(A_t S_t + \frac{B_t}{2} u_t^2 \right)$$

Where the parameters $A_t > 0$ and $B_t > 0$ are the cost coefficients, they are selected to weigh the relative importance of I_t and u_t at time step t. T is the final time. We are minimizing the number of infected individuals during the time steps $t = 0$ and $T-1$, and at the final time and also minimizing the cost of administering the control [9].

In other words, we seek the optimal control u^* such that

$$J(u^*) = \min_{u \in U} J(u)$$

Where U is the set of admissible controls defined by

$$U = \{u_t : a \leq u_t \leq b, t = 0, 1, 2, \ldots, T-1\}$$

In order to derive the necessary condition for optimal control, the pontryagin's maximum principle, in discrete time, given in was used. This principle converts into a problem of minimizing a Hamiltonian, H_t at time step t defined by

$$H_t = A_t S_t + \frac{B_t}{2} u_t^2 + \sum_{j=1}^{4} \lambda_{j,t+1} f_{j,t+1}$$

Where $f_{j,t+1}$ is the right side of the difference equation of the jth state variable at time step $t + 1$.

So the controlled mathematical system is given by the following system of difference equations.

$$P_{t+1} = P_t + pA + \beta P_t S_t - (\theta_1 + d)P_t + u_t P_t$$
$$S_{t+1} = S_t + (1 - p)A - \beta P_t S_t - (\theta + d)S_t$$
$$L_{t+1} = L_t + \theta S_t + \theta_1 P_t - (\gamma + d)L_t - u_t P_t$$

Theorem 3: Given an optimal control $u_k^* \in U$ and the solutions P_t^*, S_t^*, L_t^* of the corresponding state system (I), there exists adjoint functions $\lambda_{1,t}$, $\lambda_{2,t}$ and $\lambda_{3,t}$ satisfying

$$\lambda_{1,t} = \beta S_t(\lambda_{1,t+1} - \lambda_{2,t+1}) + \theta_1(\lambda_{3,t+1} - \lambda_{1,t+1}) - (d-1)\lambda_{1,t+1} + u_t(\lambda_{1,t+1} - \lambda_{3,t+1})$$
$$\lambda_{2,t} = A_t - (d-1)\lambda_{2,t+1} + \theta(\lambda_{3,t+1} - \lambda_{2,t+1}) + \beta P_t(\lambda_{1,t+1} - \lambda_{2,t+1})$$
$$\lambda_{3,t} = -(\gamma + d - 1)\lambda_{3,t+1}$$

With the transversality conditions at time T
$\lambda_{1,T} = \lambda_{3,T} = 0$ and $\lambda_{2,T} = A_T$
Furthermore, for $k = 0, 1, \ldots, T - 1$, the optimal control u_t^* is given by

$$u_t^* = \min\left(b, \max\left(a, \frac{\lambda_{1,t+1} - \lambda_{3,t+1}}{B_t}P_t\right)\right)$$

Proof: The Hamiltonian at time step t is given by

$$H_t = A_t S_t + \frac{B_t}{2}u_t^2 + \lambda_{1,t}\{P_t + pA + \beta P_t S_t - (\theta_1 + d)P_t + u_t P_t\}$$
$$+ \lambda_{2,t}\{S_t + (1-p)A - \beta P_t S_t - (\theta + d)S_t\}$$
$$+ \lambda_{3,t}\{L_t + \theta S_t + \theta_1 P_t - (\gamma + d)L_t - u_t P_t\}$$

For, $t = 0, 1, \ldots, T - 1$, the adjoint equations and transversality conditions can be obtained by using Pontryagin's Maximum Principle, in discrete time, given in [9] such that

$$\lambda_{1,t} = \frac{\partial H_t}{\partial P_t}, \lambda_{1,T} = 0$$

$$\lambda_{2,t} = \frac{\partial H_t}{\partial S_t}, \lambda_{2,t} = A_T$$

$$\lambda_{3,t} = \frac{\partial H_t}{\partial L_t}, \lambda_{3,t} = 0$$

For $t = 0, 1, \ldots, T - 1$, the optimal control u_t^* can be solved from the optimality condition,

$$\frac{\partial H_t}{\partial u_t} = 0$$

That is,

$$\frac{\partial H_t}{\partial u_t} = B_t u_t + (\lambda_{1,t+1} - \lambda_{3,t+1})P_t = 0$$

6 Numerical Example

According to the World Health Organisation(WHO), despite India's regulation on public smoking, 30% adults are found exposed to second hand tobacco smoke at work, the study said. WHO has also declared that there are approximately 120 million smokers in India. i.e. India is home for 12% smokers worldwide. According to National Institute of Cancer Prevention and Research in 2012, the cancer incidence among both sexes is 70,000 and the mortality rate is 64,000.

We carry out numerical simulations using MATLAB. We consider the following set of parametric values:

From these information, we have

$$x_0 = 1, \, y_0 = 1, \, z_0 = 1, \, \beta = 0.4, \, \delta = 0.9, p = 9.7, \, \gamma = 0.85$$
$$\theta = 0.04, \, d = 0.748, \, \theta_1 = 0.25, A = 100$$

Let $A = 100$, where A is the constant population entering the system. Assume x_0, y_0, z_0 denotes the initial population of non-smokers, smokers and lung cancer patients in the system. $N_0 = x_0 + y_0 + z_0 = 3$. N_0 is the total population initially. Then the constant population A enters the system and the process of contact between smokers and non-smokers begins.

The graphs given below allows us to compare changes in the number of Non smokers, smokers and lung cancer patients before and after the introduction of control (Fig. 2).

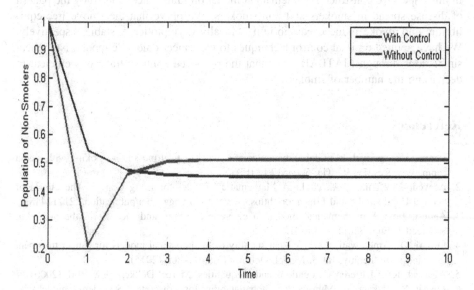

Fig. 2. Population of Non-smokers with and without control

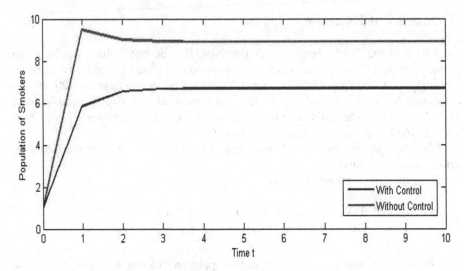

Fig. 3. Population of Smokers with and without control

In Fig. 3, we observe that there is a decrease in smokers due to control strategy.

7 Conclusion

In this paper, we construct a mathematical model on lung cancer and study the pattern of disease spread in smokers and non smokers. We prove that the smoke free equilibrium and smoke induced equilibrium is locally asymptotically stable respectively. We have applied optimal control techniques to the cancer epidemic model. Numerical simulations through MATLAB show that the proposed control strategy is efficient in decreasing the number of smokers.

References

1. Verma, V., Agarwal, M.: Global dynamics of a mathematical model on smoking with media campaigns. Res. Desk. **4**(1), 500–512 (2015)
2. Acevedo-Estefania, C.A., et.al.: A Mathematical model for Lung Cancer: The effects of Second Hand smoke and Education. https://www.researchgate.net/publication/221711448
3. Andest, J.N.: A mathematical model on cigarette smoking and nicotine in the lung. Int. Refereed J. Eng. Sci. **2**, 1–3 (2013)
4. Lestari, D., Ambarwati, R.D.: A local stability of mathematical models for cancer treatment by using gene therapy. Int. J. Model. Optim. **5**(3), 202–206 (2015)
5. Agarwal, R.P.: Difference Equations and Inequalities. Marcel Dekker, New York (2000)
6. Enatsu, Y., Nakata, Y., Muroya, Y.: Global stability for a discrete SIS epidemic model with immigration of infectives. J. Differ. Equ. Appl. **18**, 1913–1924 (2012)

7. Jang, S., Elaydi, S.: Difference equations from discretization of a continuous epidemic model with immigration of infectives. Can. Appl. Math. Q. **11**(1), 93–105 (2003)
8. Tipsri, S., Chinviriyasit, W.: Stability analysis of SEIR model with saturated incidence and time delay. Int. J. Appl. Phys. Math. **4**, (2014)
9. Sebastian, E., Victor, P.: Optimal control strategy of a discrete time SVIR Epidemic model with immigration of infectives. Int. J. Pure Appl. Math. **113**(8), 55–63 (2017)

Transient Analysis of an *M*/*M*/1 Queue with Bernoulli Vacation and Vacation Interruption

K. V. Vijayashree[1] and B. Janani[2(✉)]

[1] Department of Mathematics, College of Engineering, Anna University,
Chennai, India
`vkviji@annauniv.edu`
[2] Department of Mathematics, Vel Tech Rangarajan Dr. Sagunthala R&D
Institute of Science and Technology, Chennai, India
`jananisrini2009@gmail.com`

Abstract. This paper presents the transient analysis of an *M*/*M*/1 queueing system subject to Bernoulli vacation and vacation interruption. The arrivals are allowed to join the queue according to a Poisson distribution and the service takes place according to an exponential distribution. Whenever the system is empty, the server can take either a working vacation or an ordinary vacation with certain probabilities. During working vacation, the arrivals are allowed to join the queue and the service takes place according to an exponential distribution, but with a slower rate. The vacation times are also assumed to be exponentially distributed. Further, during working vacation, the server has the option to either continue the vacation or interrupt and transit to the regular busy period. During ordinary vacation, the arrivals are allowed to join the queue but no service takes place. The sever returns to the system on completion of the vacation duration. Upon returning from vacation, the server continues to provide service to the waiting customers (if any) or takes another vacation (working or ordinary vacation) if the system is empty. Explicit expressions are obtained for the time dependent system size probabilities in terms of modified Bessel function of the first kind using generating function and Laplace transform techniques. Numerical illustrations are added to support the theoretical results.

Keywords: System size probabilities · Transient analysis · Laplace transform
Generating function · Bernoulli vacation · Vacation interruption

1 Introduction

In recent years, queueing models subject to vacation are of interest to researchers owing to its wide spread applicability in real time situations. Server vacation may occur due to several factors like insufficient workload in human behaviour, failure of the server subject to repair period, preventive maintenance period in a production system, secondary tasks assigned to the server (which occurs in computer maintenance and testing), service rendered to arrivals as in priority queueing discipline and so on. The concept of queueing systems with server vacations was first discussed by Levy and

© Springer Nature Singapore Pte Ltd. 2018
G. Ganapathi et al. (Eds.): ICC3 2017, CCIS 844, pp. 214–228, 2018.
https://doi.org/10.1007/978-981-13-0716-4_18

Yechiali (1975). A comprehensive and detailed review of the vacation queueing model can be found in the survey by Doshi (1986) and the books by Takagi (1991) and Tian and Zhang (2006).

A classical queueing model consists of three parts: the arrival process, the service process and queue discipline. In addition, a vacation queueing model has a vacation process governed by a vacation policy. There are different types of vacation policies. The most commonly used policies are (i) Single vacation policy (ii) Multiple vacation policy and (iii) Working vacation policy. In a single vacation policy, the server takes a vacation of some random duration when there are no customers in the system. The server returns to the system at the end of the vacation and starts providing service to the waiting customer, if any, otherwise the server will wait to complete the busy period. In a multiple vacation policy, when the server returns from a vacation and finds the queue empty, he immediately takes another vacation. In a working vacation policy, the server works at a slower rate during a vacation period rather than completely stopping the service during the vacation period. Servi and Finn (2002) were the first to introduce a class of semi-vacation model called working vacation.

When the server served all the customers and there are no customers left in the system, it is eventual that the server will go for a vacation. Suppose the server have two possibilities to take a vacation, i.e., either a working vacation or an ordinary vacation with some probability, then such type of vacation policy is known as Bernoulli scheduled vacation policy. For example, consider a machine shop, with a single server. When the server is busy, items are produced continuously and if he is idle, there is no production. However, for all practical reasons, the server might either take a vacation of some random duration with probability p or decide to provide service at a reduced rate with probability, $(1-p)$. Further, by offering service at a reduced rate, the server may continue to do so with probability, q or due to certain unforeseen reasons, like a sudden increase in the demand, interrupt the vacation with probability $(1-q)$, and continue the busy period. The demands are assumed to vary from time to time with the rate which is independent of the state of the server. The level of inventory thus oscillates, depending on the busy or idle state of the server. Such scenario can be modelled as a queue subject to Bernoulli vacation and vacation interruption.

The classical vacation scheme with Bernoulli schedule in which the server serves the new customer with probability p or takes a vacation with probability $(1-p)$ was originated and developed significantly by Keilson and Servi (1986). The advantage of the Bernoulli schedule is the existence of a control parameter p. By adjusting the value of p, we can control the congestion of the system. Various aspects of Bernoulli vacation models for single server queueing systems have been studied by Servi and Finn (2002), Ramaswamy and Servi (1988), Tadjet al (2006) among several others. Vacation queueing systems with vacation interruption were first introduced by Li and Tian in (2007) where they studied the $M/M/1$ queue with working vacation and vacation interruption. Zhang and Shi (2009) derived the stationary probabilities of an $M/M/1$ queue subject to Bernoulli-Controlled-Scheduled vacation and vacation interruption. Analysis of $M/G/1$ unreliable server queue with two phases of service and Bernoulli vacation schedule was done by Choudhury and Deka (2012). Goswami (2014) provided stationary probabilities of customers' impatience in Markovian queueing system with multiple working vacations and Bernoulli schedule vacation interruption.

Bouchentouf and Yahiaoui (2017) provided stationary analysis of feedback queueing system with reneging and retention of reneged customers, multiple working vacations and Bernoulli schedule vacation interruption.

The above mentioned contributions on working vacation are confined to the analysis in the stationary regime. The study of the transient solution of queueing systems taking together the above mentioned features is interesting, however due to the complexity of the problem not much work in this direction is addressed in the literature. This paper deals with the transient solution of an $M/M/1$ queueing model with Bernoulli vacation and vacation Interruption. The paper is organized as follows: Sect. 2 deals with the description of the model under consideration by introducing the various parameters and other notations. Section 3 presents the generating function approach to obtain the transient system size probabilities. Section 4 presents the corresponding stationary solution as a limiting case of the transient solution. Section 5 depicts the numerical illustrations of the behaviour of various system size probabilities against different values of the influencing parameters.

2 Model Description

Consider a single server queueing model wherein customers arrive according to a Poisson distribution with parameter, λ and are serviced according to an exponential distribution with parameter, μ. When the system becomes empty, the server has the option to take either an ordinary vacation with probability, $p(0 \leq p \leq 1)$ or a working vacation with probability $(1 - p)$. Both the ordinary and working vacation times are assumed to follow exponential distribution with parameters θ and θ_v respectively. During ordinary vacation the customers continue to arrive and join the queue, however no service takes place. During the working vacation times, the server is assumed to provide service according to an exponential distribution with a slower service rate, $\mu_v(<\mu)$. Further, when the server is in working vacation, after completion of each service, he can either continue the vacation with probability q or interrupt the vacation with probability $(1 - q)$ and resume to regular busy period.

Let $X(t)$ denotes the total number of customers in the system and $J(t)$ represents the state of the server at time t. The state $J(t) = 1$ refers to the server being in functional state, $J(t) = 0$ refers to the server being in ordinary vacation state and $J(t) = 2$ refers to the server being in working vacation state at time t. Then $(J(t), X(t))$ defines a two-dimensional continuous time Markov process with state space, $S = \{(0,0)\} \cup \{(2,0)\} \cup \{(j,n)\}; n = 1, 2 \ldots; j = 0, 1, 2\}$. The state transition diagram for the model is given in Fig. 1.

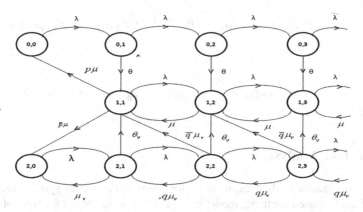

Fig. 1. State Transition Diagram of an *M/M/1* queueing system subject to Bernoulli Vacation and Vacation interruption

Let $P_{j,n}(t)$ denote the time dependent probability for the system to be in state j with n customers at time t. Mathematically

$$P_{0,0}(t) = P(J(t) = 0, X(t) = 0),$$
$$P_{2,0}(t) = P(J(t) = 2, X(t) = 0),$$

and

$$P_{j,n}(t) = P(J(t) = j, X(t) = n), j = 0, 1, 2; n = 1, 2, 3. \ldots$$

Assume that initially the system is empty and the server is in working vacation i.e. $P_{2,0}(0) = 1$. Let $\bar{p}$ and $\bar{q}$ represent the quantity $1 - p$ and $1 - q$ respectively. By standard methods, the system of Kolmogorov differential difference equations governing the process are given by

$$P'_{2,0}(t) = -\lambda P_{2,0}(t) + \mu_v P_{2,1}(t) + \bar{p}\mu P_{1,1}(t), \tag{2.1}$$

$$P'_{2,1}(t) = -(\lambda + \mu_v + \theta_v)P_{2,1}(t) + \lambda P_{2,0}(t) + q\mu_v P_{2,2}(t), \tag{2.2}$$

$$P'_{2,n}(t) = -(\lambda + q\mu_v + \theta_v + \bar{q}\mu_v)P_{2,n}(t) + \lambda P_{2,n-1}(t) + q\mu_v P_{2,n+1}(t), n = 2, 3 \ldots \tag{2.3}$$

$$P'_{00}(t) = -\lambda P_{0,0}(t) + p\mu P_{1,1}(t), \tag{2.4}$$

$$P'_{0,n}(t) = -(\lambda + \theta)P_{0,n}(t) + \lambda P_{0,n-1}(t), n = 1, 2, 3. \ldots \tag{2.5}$$

$$P'_{1,1}(t) = -(\lambda + p\mu + \bar{p}\mu)P_{1,1}(t) + \theta_v P_{2,1}(t) + \mu P_{1,2}(t) + \bar{q}\mu_v P_{2,2}(t) + \theta P_{0,1}(t), \tag{2.6}$$

and

$$P'_{1,n}(t) = -(\lambda+\mu)P_{1,n}(t) + \mu P_{1,n+1}(t) + \lambda P_{1,n-1}(t) + \theta_v P_{2,n}(t) + \bar{q}\mu_v P_{2,n+1}(t) + \theta P_{0,n}(t),$$

$$n = 2,3.\ldots$$

$$(2.7)$$

subject to the conditions $P_{2,0}(0) = 1, P_{0,0}(0) = 0$ and $P_{j,n}(0) = 0$ for $j = 0, 1, 2$ and $n = 1, 2, 3, \ldots$.

3 Transient Analysis

This section presents explicit expressions for the time dependent system size probabilities of the proposed queueing model in terms of modified Bessel function of the first kind using Laplace transform and generating function methodology.

3.1 Evaluation of $P_{2,n}(t)$

Let $\hat{P}_{j,n}(s)$ denote the Laplace transform of $P_{j,n}(t)$ for all possible values of 'j' and 'n'. Define the generating function,

$$Q(z,t) = \sum_{n=1}^{\infty} P_{2,n}(t)z^n.$$

Then

$$Q'(z,t) = \sum_{n=1}^{\infty} P'_{2,n}(t)z^n.$$

Multiplying the Eq. (2.3) by z^n and summing it over all values of 'n' leads to

$$Q(z,t) = \lambda z \int_0^t P_{2,0}(y)e^{\left(\lambda z + \frac{q\mu_v}{z}\right)(t-y)}e^{-(\lambda+\theta_v+\mu_v)(t-y)}dy$$

$$- q\mu_v \int_0^t P_{2,1}(y)e^{\left(\lambda z + \frac{q\mu_v}{z}\right)(t-y)}e^{-(\lambda+\theta_v+\mu_v)(t-y)}dy.$$

It is well known that if $\alpha_1 = 2\sqrt{\lambda q\mu_v}$ and $\beta_1 = \sqrt{\frac{\lambda}{q\mu_v}}$ then $e^{\left(\lambda z + \frac{q\mu_v}{z}\right)} = \sum_{n=-\infty}^{\infty}(\beta_1 z)^n I_n(\alpha_1(t))$. Therefore, $Q(z,t)$ can be expressed as

$$Q(z,t) = \lambda z \int_0^t P_{2,0}(y)e^{-(\lambda+\theta_v+\mu_v)(t-y)}\sum_{n=-\infty}^{\infty}(\beta_1 z)^n I_n(\alpha_1(t-y))dy$$

$$- q\mu_v \int_0^t P_{2,1}(y)e^{-(\lambda+\theta_v+\mu_v)(t-y)}\sum_{n=-\infty}^{\infty}(\beta_1 z)^n I_n(\alpha_1(t-y))dy.$$

$$(3.1)$$

Comparing the coefficients of z^n on both sides of Eq. (3.1) leads to

$$P_{2,n}(t) = \lambda \int_0^t P_{2,0}(y)e^{-(\lambda+\theta_v+\mu_v)(t-y)}\beta_1^{n-1}I_{n-1}(\alpha_1(t-y))dy$$
$$- q\mu_v \int_0^t P_{2,1}(y)e^{-(\lambda+\theta_v+\mu_v)(t-y)}\beta_1^n I_n(\alpha_1(t-y))dy. \tag{3.2}$$

Similarly, comparing the coefficients of z^{-n} on both sides of Eq. (3.1) yields

$$0 = \lambda \int_0^t P_{2,0}(y)e^{-(\lambda+\theta_v+\mu_v)(t-y)}\beta_1^{n-1}I_{n+1}(\alpha_1(t-y))dy$$
$$- q\mu_v \int_0^t P_{2,1}(y)e^{-(\lambda+\theta_v+\mu_v)(t-y)}\beta_1^n I_n(\alpha_1(t-y))dy, \tag{3.3}$$

Subtracting Eq. (3.3) from Eq. (3.2), we get

$$P_{2,n}(t) = \lambda \int_0^t P_{2,0}(y)e^{-(\lambda+\theta_v+\mu_v)(t-y)}\beta_1^{n-1}\frac{2nI_n(\alpha_1(t-y))}{\alpha_1(t-y)}dy, n = 1, 2\ldots, \tag{3.4}$$

which can be rewritten as

$$P_{2,n}(t) = 2\lambda(\alpha_1\beta_1)^{n-1}\int_0^t P_{2,0}(y)e^{-(\lambda+\theta_v+\mu_v)(t-y)}\frac{nI_n(\alpha_1(t-y))}{\alpha_1^n(t-y)}dy. n = 1, 2, \ldots. \tag{3.5}$$

Thus the transient state probabilities during the working vacation state of the server is determined in terms of $P_{2,0}(t)$ which is yet to be determined. Towards this end, taking Laplace transform of the above equation leads to

$$\hat{P}_{2,n}(s) = 2\lambda(\alpha_1\beta_1)^{n-1}\hat{P}_{2,0}(s)\left(\omega_1 + \sqrt{\omega_1^2 - \alpha^2}\right)^{-n}, \tag{3.6}$$

Where $\omega_1 = s + \lambda + \theta_v + \mu_v$. Substituting $n = 1$ in Eq. (3.6) leads to

$$\hat{P}_{2,1}(s) = 2\lambda\hat{P}_{2,0}(s)\left(\omega_1 + \sqrt{\omega_1^2 - \alpha^2}\right)^{-1}. \tag{3.7}$$

Also, taking Laplace transform of Eq. (2.1) leads to

$$\hat{P}_{2,0}(s) = \frac{1}{s+\lambda} + \frac{\mu_v}{s+\lambda}\hat{P}_{2,1}(s) + \frac{\bar{p}\mu}{s+\lambda}\hat{P}_{1,1}(s).$$

Substituting Eq. (3.7) in the above equation, we get

$$\hat{P}_{2,0}(s) = \frac{1}{s+\lambda} + \frac{\mu_v}{s+\lambda}\left(2\lambda\hat{P}_{2,0}(s)\left(\omega_1 + \sqrt{\omega_1^2 - \alpha^2}\right)^{-1}\right) + \frac{\bar{p}\mu}{s+\lambda}\hat{P}_{1,1}(s),$$

which can also be expressed as

$$\hat{P}_{2,0}(s) = \frac{1}{1 - \left(\frac{2\lambda\mu_v}{(s+\lambda)\left(\omega_1 + \sqrt{\omega_1^2 - \alpha^2}\right)}\right)}\left(\frac{1}{s+\lambda} + \frac{\bar{p}\mu}{s+\lambda}\hat{P}_{1,1}(s)\right).$$

Hence,

$$\hat{P}_{2,0}(s) = \left(\frac{1}{s+\lambda} + \frac{\bar{p}\mu}{s+\lambda}\hat{P}_{1,1}(s)\right)\sum_{k=0}^{\infty}\sigma^k(s), \qquad (3.8)$$

Where

$$\sigma(s) = \frac{2\lambda\mu_v}{(s+\lambda)\left(\omega_1 + \sqrt{\omega_1^2 + \alpha^2}\right)}.$$

Taking inverse Laplace transform for the above equation, leads to

$$P_{2,0}(t) = \left(e^{-\lambda t} + \bar{p}\mu e^{-\lambda t} * P_{1,1}(t)\right) * \sum_{k=0}^{\infty}\sigma_1(t),$$

Where

$$\sigma_1(t) = L^{-1}\left(\sigma^k(s)\right) = L^{-1}\left(\frac{2\lambda\mu_v}{(s+\lambda)\left(\omega_1 + \sqrt{\omega_1^2 + \alpha^2}\right)}\right)$$

$$= (2\lambda\mu_v)^k e^{-\lambda t}\frac{(\lambda t)^{k-1}}{(k-1)!} * e^{-(\lambda+\theta_v+\mu_v)t}\frac{I_k(\alpha_1 t)}{\alpha_1 t}.$$

Substituting for $\hat{P}_{2,0}(s)$ from Eq. (3.8) in Eq. (3.6) yields

$$\hat{P}_{2,n}(s) = 2\lambda(\alpha_1\beta_1)^{n-1}\left(\omega_1 + \sqrt{\omega_1^2 + \alpha^2}\right)^{-n}\left(\frac{1}{s+\lambda} + \frac{\bar{p}\mu}{s+\lambda}\hat{P}_{1,1}(s)\right)\sum_{k=0}^{\infty}\sigma^k(s),$$

which on inversion leads to

$$P_{2,n}(t) = 2\lambda(\alpha_1\beta_1)^{n-1}e^{-(\lambda+\theta_v+\mu_v)t}\left(\frac{nI_n(\alpha_1 t)}{\alpha_1^n t}\right) * \left(e^{-\lambda t} + \bar{p}\mu e^{-\lambda t}P_{1,1}(t)\right) * \sum_{k=0}^{\infty}\sigma_1(t).$$

3.2   Evaluation of $P_{0,n}(t)$

Taking Laplace transform of Eq. (2.5) leads to

$$\hat{P}_{0,n}(s) = \frac{\lambda}{s+\lambda+\theta}\hat{P}_{0,n-1}(s); n = 1, 2\ldots,$$

which yields

$$\hat{P}_{0,n}(s) = \left(\frac{\lambda}{s+\lambda+\theta}\right)^n \hat{P}_{0,0}(s); n = 1, 2\ldots \tag{3.9}$$

Similarly, taking Laplace transform of Eq. (2.4) yields

$$\hat{P}_{0,0}(s) = \frac{p\mu}{s+\lambda}\hat{P}_{1,1}(s). \tag{3.10}$$

Substituting Eq. (3.10) in Eq. (3.9), we get

$$\hat{P}_{0,n}(s) = \left(\frac{\lambda}{s+\lambda+\theta}\right)^n \left(\frac{p\mu}{s+\lambda}\right)\hat{P}_{1,1}(s); n = 0, 1\ldots \tag{3.11}$$

which on inversion leads to

$$P_{0,n}(t) = \lambda^n \frac{t^{n-1}e^{-(\lambda+\theta)t}}{(n-1)!} * p\mu e^{-\lambda t} * P_{1,1}(t). \tag{3.12}$$

Thus all the transient state probabilities during the vacation state of the server is expressed in terms of $P_{1,1}(t)$. It still remains to determine the transient probabilities during functional state of the server represented by $P_{1,n}(t)$.

3.3   Evaluation of $P_{1,n}(t)$

Define

$$H(z,t) = \sum_{n=1}^{\infty} P_{1,n}(t)z^n.$$

Then

$$H'(z,t) = \sum_{n=1}^{\infty} P'_{1,n}(t)z^n.$$

Multiplying Eq. (2.7) by z^n and summing it over all values of 'n' leads to

$$H'(z,t) - \left(\frac{\mu}{z} + \lambda z - (\lambda + \mu)\right)H(z,t) = \left(\theta_v + \frac{\bar{q}\mu_v}{z}\right)Q(z,t) - \mu P_{1,1}(t)$$
$$- \bar{q}\mu_v P_{2,1}(t) + \theta \sum_{n=1}^{\infty} P_{0,n}(t)z^n,$$

On integrating with respect to t, we get

$$H(z,t) = \int_0^t \left\{ \left(\theta_v + \frac{\bar{q}\mu_v}{z}\right)Q(z,y) - \mu P_{1,1}(y) - \bar{q}\mu_v P_{2,1}(y) \right.$$
$$\left. + \theta \sum_{n=1}^{\infty} P_{0,n}(y)z^n \right\} e^{\left(\frac{\mu}{z} + \lambda z - (\lambda + \mu)\right)(t-y)} dy. \tag{3.13}$$

Using

$$exp\left(\frac{\mu}{z} + \lambda z\right)t = \sum_{n=-\infty}^{\infty} (\beta z)^n I_n(\alpha t),$$

and comparing the coefficients of z^n on both sides of Eq. (3.13), we get

$$P_{1,n}(t) = \theta_v \int_0^t e^{-(\lambda+\mu)(t-y)} \sum_{k=0}^{\infty} P_{2,n+k}(y)\beta^{-k} I_k(\alpha(t-y))dy$$
$$+ \bar{q}\mu_v \int_0^t e^{-(\lambda+\mu)(t-y)} \sum_{k=n}^{\infty} P_{2,n-(k-1)}(y)\beta^k I_k(\alpha(t-y))dy$$
$$- \mu \int_0^t P_{1,1}(y)e^{-(\lambda+\mu)(t-y)} \beta^n I_n(\alpha(t-y))dy \tag{3.14}$$
$$- \bar{q}\mu_v \int_0^t P_{2,1}(y)e^{-(\lambda+\mu)(t-y)} \beta^n I_n(\alpha(t-y))dy$$
$$+ \theta \int_0^t e^{-(\lambda+\mu)(t-y)} \sum_{k=0}^{\infty} P_{0,n+k}(y)\beta^{-k} I_k(\alpha(t-y))dy.$$

It is seen that from Eq. (3.14) that $P_{1,n}(t)$ is expressed in terms of $P_{2,n}(t), P_{0,n}(t)$ and $P_{1,1}(t)$. Also, $P_{2,n}(t)$ and $P_{0,n}(t)$ are already in terms of $P_{1,1}(t)$.

Hence we have expressed all the transient state probabilities during working vacation and ordinary vacation state, namely, $P_{2,n}(t)$ and $P_{0,n}(t)$ in terms of $P_{1,1}(t)$. Therefore, all the time dependent probabilities are in terms of $P_{1,1}(t)$. From the normalization condition given by

$$\sum_{n=0}^{\infty} P_{2,n}(t) + \sum_{n=0}^{\infty} P_{0,n}(t) + \sum_{n=1}^{\infty} P_{1,n}(t) = 1,$$

the quantity $P_{1,1}(t)$ can be obtained in closed form.

4 Stationary Analysis

Let $\pi_{j,n}$ denote the steady state probability for the system to be in state j with n customers. Mathematically,

$$\pi_{j,n} = \lim_{t \to \infty} P_{j,n}(t).$$

Using the final value theorem of Laplace transform, it is observed that $\pi_{j,n} = \lim_{s \to 0} s \hat{P}_{j,n}(s)$. Therefore, from Eq. (3.8), we get

$$\lim_{s \to 0} s \hat{P}_{2,0}(s) = \lim_{s \to 0} \left(\frac{s}{s+\lambda} + \frac{\bar{p}\mu}{s+\lambda} s \hat{P}_{1,1}(s) \right) \sum_{k=0}^{\infty} \sigma^k(s).$$

On substituting for $\sigma^k(s)$ and after some algebra, we obtain

$$\pi_{2,0} = \left(\frac{\bar{p}\mu}{\lambda} \pi_{1,1} \right) \sum_{k=0}^{\infty} \left(\frac{2\mu_v}{(\lambda + \theta_v + \mu_v) + \sqrt{(\lambda + \theta_v + \mu_v)^2 - \alpha^2}} \right)^k,$$

which on further simplification yields

$$\pi_{2,0} = \left(\frac{\bar{p}\mu}{\lambda} \pi_{1,1} \right) \left(\frac{1}{1 - \frac{2\mu_v \left[(\lambda + \theta_v + \mu_v) + \sqrt{(\lambda + \theta_v + \mu_v)^2 - \alpha^2} \right]}{4\lambda q \mu_v}} \right). \tag{4.1}$$

Similarly, from Eq. (3.9), we get

$$\lim_{s \to 0} s \hat{P}_{0,n}(s) = \lim_{s \to 0} \left(\frac{\lambda}{s+\lambda+\theta} \right)^n \left(\frac{p\mu}{s+\lambda} \right) s \hat{P}_{1,1}(s); n = 0, 1. \ldots$$

which on further simplification leads to

$$\pi_{0,n} = \left(\frac{\lambda}{\lambda + \theta} \right)^n \left(\frac{p\mu}{\lambda} \right) \pi_{1,1} \tag{4.2}$$

By taking Laplace transform of Eq. (3.14) and multiplying both sides by s, and taking $\lim_{s \to 0}$ on both sides, we get

$$\lim_{s\to0} s\hat{P}_{1,n}(s) = \lim_{s\to0}\left\{\theta_v\sum_{k=0}^{\infty}\beta^{-k}s\hat{P}_{2,n+k}(s)\frac{(\omega_1-\sqrt{\omega_1^2-\alpha_2})^k}{\alpha^k\sqrt{\omega_1^2-\alpha^2}}\right.$$

$$+\bar{q}\mu_v\sum_{k=n}^{\infty}\beta^k s\hat{P}_{2,n-(k-1)}(s)\frac{(\omega_1-\sqrt{\omega_1-\sqrt{\omega_1^2-\alpha_2}})^k}{\alpha^k\sqrt{\omega_1^2-\alpha^2}}-\mu\beta^2 s\hat{P}_{1,1}(s)\frac{(\omega_1-\sqrt{\omega_1^2-\alpha^2})^n}{\alpha^n\sqrt{\omega_1^2-\alpha^2}}$$

$$\left.-\bar{q}\mu_v\beta^n s\hat{P}_{2,1}(s)\frac{(\omega_1-\sqrt{\omega_1^2-\alpha^2})^n}{\alpha^n\sqrt{\omega_1^2-\alpha^2}}+\theta\sum_{k=n}^{\infty}\beta^{-k}s\hat{P}_{0,n+k}(s)\frac{(\omega_1-\sqrt{\omega_1-\sqrt{\omega_1^2-\alpha_2}})^k}{\alpha^k\sqrt{\omega_1^2-\alpha^2}}\right\},$$

which on further simplification leads to

$$\pi_{1,n} = \pi_{1,1}\sum_{k=0}^{n-1}\left(\frac{\lambda}{\mu}\right)^{n-k-1}\left[p\left(\frac{\lambda}{\lambda+\theta}\right)^k\right.$$

$$\left.+(1-p)\left(\frac{(\lambda+\theta_v+\mu_v)+\sqrt{(\lambda+\theta_v+\mu_v)^2-4\lambda q\mu_v}}{2q\mu_v}\right)^k\right]. \tag{4.3}$$

The steady state probabilities given by Eqs. (4.1), (4.2) and (4.3) are seen to coincide with the probabilities obtained by Zhang and Shi (2009) subject to the stability condition, $\frac{\lambda}{\mu}<1$.

5 Numerical Illustrations

This section illustrates the behaviour of time dependent state probabilities of the system during the functional state and vacation state of the server against time for appropriate choice of the parameter values. Though the system is of infinite capacity, the value of N is restricted to 25 for the purpose of numerical study.

Figure 2 depicts the variation of the system size probability, $P_{2,0}(t)$ against time for $\lambda = 0.25, \mu = 2, \theta = 0.3, \theta_v = 0.7, \mu_v = 1.5$ and varying values of p (namely 0.1, 0.3, 0.5 and 0.8). By our assumption, the system is initially assumed to be empty and the server is in working vacation. Therefore, the graph of $P_{2,0}(t)$ starts at 1 and converges to the corresponding steady state probabilities as time progresses. It is seen that for a fixed time t, the probability, $P_{2,0}(t)$ increases with decreases in p. Observe that an increase in the value of p will make the system to be in ordinary vacation state with more probability as seen in the state transition diagram.

Figure 3 depicts the variation of the system size probability $P_{0,n}(t)$ against t for $\lambda = 0.25, \mu = 2, \theta = 0.3, \theta_v = 0.7, \mu_v = 1.5, p = 0.5, q = 0.5$ and varying values of 'n'. It is seen that at an arbitrary epoch of time, $P_{0,n}(t)$ decreases with increase in 'n' and converges to the corresponding steady state probability as time progresses. For the choice of λ, it is more likely to find the system with less number of customers initially and since there are no service during ordinary vacation duration the probability for the corresponding number of customers tend to increase as time progresses. Further, since

θ is taken to be 0.3, it is unlikely to find the system with more customers during ordinary vacation duration at any instant of time t.

Figure 4 depicts the variation of the system size probability $P_{1,n}(t)$ against t for $\lambda = 0.25$, $\mu = 2$, $\theta = 0.3$, $\theta_v = 0.7$, $\mu_v = 1.5$, $p = 0.5$, $q = 0.5$ and varying values of 'n'. It is seen that at an arbitrary epoch of time, $P_{1,n}(t)$ decreases with increase in'n' and converges to the corresponding steady state probabilities as time progresses. Note that when the server is in the functional state, the arrivals are continuously serviced at an exponential rate and based on the values of θ and θ_v the transition from the vacation state to the functional state of the server takes place at various instants of time. However, it is more unlikely to find the system with more number of customers at any instant of time for the given choice of the parameter values.

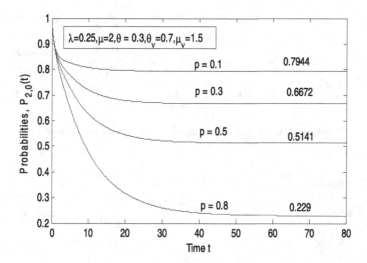

Fig. 2. Variation of $P_{2,0}(t)$ against t for varying values of p

Tables 1 and 2 depicts the comparison of the probabilities for the server to be in busy state and ordinary vacation state respectively for the same choice of parameter values and varying values of$n$ at various instants of time, t. Observe that at any instant of time the normalization condition is satisfied and that as time progresses the last row of the table depicts the convergence of the transient probabilities to the corresponding steady state values.

Table 2 depicts the comparison of the probabilities for the server to be in ordinary vacation state for the same choice of parameter values and varying values of n at various instants of time, t.

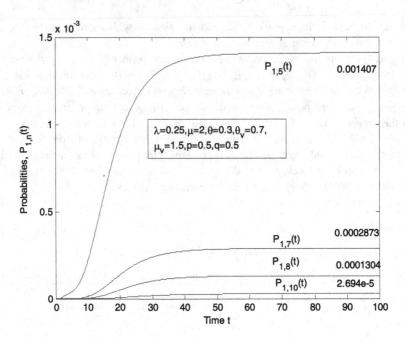

Fig. 3. Variation of $P_{1,n}(t)$ against t for varying values of n

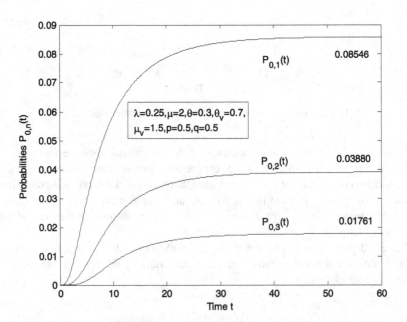

Fig. 4. Variation of $P_{0,n}(t)$ against t for varying values of n

Table 1. Asymptotic convergence of $P_{1,n}(t)$

t	$P_{1,5}(t)$	$P_{1,7}(t)$	$P_{1,8}(t)$	$P_{1,10}(t)$
0	0	0	0	0
4	0.00002911266890	0.00000037444308	0.00000003739969	0.00000000029576
8	0.00011481845956	0.00000425214852	0.00000074452411	0.00000001857952
12	0.00035910133255	0.00002820576324	0.00000722586369	0.00000038887096
16	0.00066752385098	0.00007860290621	0.00002511278947	0.00000216889102
20	0.00092834022263	0.00013793064051	0.00005057154785	0.00000597234117
24	0.00111074681932	0.00018883541422	0.00007544507728	0.00001096997375
28	0.00122747445357	0.00022548851528	0.00009494022993	0.00001580657215
32	0.00129928819410	0.00024949351461	0.00010838623844	0.00001965722485
36	0.00134275620463	0.00026447097309	0.00011702301660	0.00002236828404
40	0.00136890202218	0.00027360247445	0.00012236795982	0.00002413988135
44	0.00138459273210	0.00027911320974	0.00012561636357	0.00002524911472
48	0.00139400165003	0.00028242488338	0.00012757449897	0.00002592788691
52	0.00139964221229	0.00028441178255	0.00012875079046	0.00002633851811
56	0.00140302338914	0.00028560313983	0.00012945644102	0.00002658561081
60	0.00140505014353	0.00028631733090	0.00012987953492	0.00002673394767
64	0.00140626501369	0.00028674543872	0.00013013316394	0.00002682291113
68	0.00140699322340	0.00028700205197	0.00013028519385	0.00002687624481
72	0.00140742972090	0.00028715586732	0.00013037632076	0.00002690821342
76	0.00140769136213	0.00028724806459	0.00013043094176	0.00002692737457
80	0.00140784819247	0.00028730332778	0.00013046368126	0.00002693885912
84	0.00140794219806	0.00028733645269	0.00013048330517	0.00002694574261

Table 2. Asymptotic convergence of $P_{0,n}(t)$)

t	$P_{0,1}(t)$	$P_{0,2}(t)$	$P_{0,3}(t)$
0	0	0	0
4	0.02042140317923	0.00429527157994	0.00075585480875
8	0.04929610669518	0.01699050041951	0.00525746177526
12	0.06522028504807	0.02643130320382	0.01015991159029
16	0.07378535032242	0.03176146830920	0.01334853847433
20	0.07866291497186	0.03474317677038	0.01517498095393
24	0.08153221046054	0.03646641332349	0.01621869344860
28	0.08324153061707	0.03748412133772	0.01682786701145
32	0.08426415456303	0.03809092359566	0.01718880179242
36	0.08487677434871	0.03845401026322	0.01740419754432
40	0.08524392490622	0.03867152855365	0.01753310914398
44	0.08546398985209	0.03880189012281	0.01761034105211
48	0.08559589855623	0.03888002697182	0.01765662751171
52	0.08567496655248	0.03892686280503	0.01768437097478
56	0.08572236120809	0.03895493687829	0.01770100062012
60	0.08575077036506	0.03897176493654	0.01771096866765

6 Conclusion and Future Scope

This paper analyses a single server queueing model subject to Bernoulli vacation and vacation interruption in the transient regime. Explicit analytical expressions for the time dependent state probabilities during both functional and vacation states of the server are presented using Laplace transform and generating function methodologies in terms of modified Bessel function of the first kind. The corresponding steady state probabilities are deduced using the final value theorem of Laplace transforms and are shown to coincide with the expression obtained by Zhang and Shi (2009). For certain specific values of the parameters, numerical illustrations are added to depict the variations of the state probabilities. The convergence of the probabilities to the corresponding steady state values are also illustrated in the respective figures. Further, the model can be extended for a finite capacity system or a multi server system.

References

Bouchentouf, A.A., Yahiaoui, L.: On feedback queueing system with reneging and retention of reneged customers, multiple working vacations and Bernoulli schedule vacation interruption. Arab. J. Math. **6**, 1–11 (2017)

Doshi, B.T.: Queueing systems with vacations - a survey. Queueing Syst. Theory Appl. **1**, 29–66 (1986)

Choudhury, G., Deka, M.: A single server queueing system with two phases of service subject to server breakdown and Bernoulli vacation. Appl. Math. Model. **36**, 6050–6060 (2012)

Keilson, J., Servi, L.D.: Oscillating random walk models for $GI/G/1$ vacation systems with Bernoulli schedules. J. Appl. Prob. **23**, 790–802 (1986)

Ramaswami, R., Servi, L.D.: The busy period of the $M/G/1$ vacation model with a Bernoulli schedule. Commun. Stat. Stochast. Models **4**, 507–521 (1988)

Takagi, H.: Queueing Analysis: A Foundation of Performance Analysis, Volume 1: Vacation and Priority Systems, Part 1. Elsevier, Amsterdam (1991)

Levy, Y., Yechiali, U.: Utilization of idle time in an $M/G/1$ queueing system. Manage. Sci. **22**, 202–211 (1975)

Li, J., Tian, N.: The $M/M/1$ queue with working vacations and vacation interruption. J. Syst. Sci. Syst. Eng. **16**, 121–127 (2007)

Servi, L.D., Finn, S.G.: $M/M/1$ queues with working vacations (M/M/1/WV). Perform. Eval. **50**, 41–52 (2002)

Tadj, L., Choudhury, G., Tadj, C.: A quorum system with a random setup under N-policy and with Bernoulli vacation schedule. Qual. Technol. Quant. Manage. **3**, 145–160 (2006)

Tian, N., Zhang, G.: Vacation Queueing Models: Theory and Applications. Springer, New York (2006)

Goswami, V.: Analysis of impatient customers in queues with Bernoulli schedule working vacations and vacation interruption. J. Stochast. **2014**, 1–10 (2014)

Zhang, H., Shi, D.: The $M/M/1$ queue with Bernoulli-schedule-controlled-vacation and vacation interruption. Int. J. Inf. Manage. Sci. **20**, 579–587 (2009)

Behavioral Analysis of Temperature Distribution in a Three Fluid Heat Exchanger via Fractional Derivative

Bagyalakshmi Morachan[✉], Sai Sundara Krishnan Gangadharan, and Madhu Ganesh

PSG College of Technology, Coimbatore 641004, India
bagyalakshmisankar@gmail.com, g_ssk@yahoo.com, madhu@asme.org

Abstract. This paper analyzes the energy transfer in a three fluid cross flow plate fin type of heat exchanger using fractional differential equations and solves a fractional mathematical model governing the heat transfer process using fractional differential transform method. Thus an analytical solution is obtained, through which the anomalous behavior in the temperature distributions is studied for various fractional values. The difference in the temperature distribution pattern for various fractional parameter is illustrated graphically and their difference in the mean temperature distribution is statistically validated. The convergence of the series solution is also verified and the results are presented. The proposed analytical solution shows good agreement with the solution obtained using finite difference method.

Keywords: Fractional derivative · Heat transfer
Fractional differential transform method

1 Introduction

In recent years, fractional differential equations have gained high priority in the study of various physical problems in science and engineering due to the physical importance of memory property of the fractional derivative. Several analytical and numerical methods have been proposed for solving fractional differential equations. Shrivastava et al. (2004) analytically studied the effect of single and combined non-dimensional design parameters on the temperature distributions of the different fluid streams in a two and three-fluid heat exchangers. Mishra and Sahoo (2010) numerically investigated the transient response in the temperature distribution of a three-fluid heat exchangers due to the nonuniformities present in the inlet temperature. Goyal et al. (2014) have analyzed the multistream plate fin heat exchangers using finite volume method. Zhua et al. (2015) applied the method of decoupling transformations to analyze the performance of one dimensional multistream parallel-plate fin heat exchanger. Liu et al. (2015) developed a stable two-grid algorithm based on the mixed finite element method to solve nonlinear fourth-order time fractional reaction diffusion equation. Saraireh (2016) have analyzed

© Springer Nature Singapore Pte Ltd. 2018
G. Ganapathi et al. (Eds.): ICC3 2017, CCIS 844, pp. 229–241, 2018.
https://doi.org/10.1007/978-981-13-0716-4_19

both steady and transient state behavior of a counter flow plate fin type of heat exchanger through CFD simulation and predicted the temperature distribution and exit temperatures. Chen et al. (2016) proposed a finite difference scheme in time and spectral approximation using Laguerre functions in space to solve time fractional reaction diffusion equation on a semi-infinite spatial domain. Feng et al. (2016) applied different numerical scheme for the time derivative and finite element method for the space to analyze the space-time fractional diffusion equations on a finite domain. Kukla and Seidlecka (2017) have analyzed the heat transfer in a composite sphere through time fractional heat conduction and provided an analytical solution to the problem for time dependent surrounding temperature and for time space dependent volumetric heat source. Mostly, the authors have used discretization techniques to solve engineering problems, which are computationally costly.

This paper proposes the fractional differential transform method to analytically study the temperature distribution in a single pass cross flow three fluid plate fin heat exchanger through a system of fractional energy equations. The anomalous behavior of the heat transfer process is analyzed for various fractional values and the results are illustrated graphically using both the surface and contour plots. The mean difference in the temperature distribution pattern for various fractional values are statistically validated. Such anomalous behavior in the temperature distribution for different fractional values cannot be obtained using usual integer order energy balance equations. Thus, the fractional mathematical model plays an important role in the heat transfer analysis.

Section 2 presents the required definitions of fractional calculus and the fractional differential transform method. Section 3 describes the fractional mathematical model and the solution methodology is presented in Sect. 4. Section 5 briefly discusses the analytical results and the conclusion is given in Sect. 6.

2   Preliminaries

This section presents the most frequently used definitions of fractional derivative of order $\alpha > 0$ (Podlubny 1999) and the basic properties of DTM used in this article.

Definition 2.1. The Riemann Liouville fractional derivative of order α is defined as

$$D_{x_0}^{\alpha} f(x) = \frac{1}{\Gamma(m-\alpha)} \left[\frac{d^m}{dx^m} \int_{x_0}^{x} (x-\xi)^{m-\alpha-1} f(\xi) d\xi \right], \quad \alpha > 0, x > 0 \quad (1)$$

Definition 2.2. The Caputo fractional derivative of order α is defined as

$$D_{x_0}^{\alpha} f(x) = \frac{1}{\Gamma(m-\alpha)} \int_{x_0}^{x} \frac{f^{(m)}(\xi)}{(x-\xi)^{\alpha-m+1}} d\xi, \quad \alpha > 0, x > 0 \quad (2)$$

where $m - 1 < \alpha \leq m, \ m \in N$.

Definition 2.3. The author Zhou (1986) introduced differential transform method in solving electrical circuit problems. The generalized two-dimensional fractional differential transform [Bildik et al. 2006; Erturk and Momani 2008] is given below:

Consider a function of two variables $u(x,y)$ represented as a product of two single-variable functions $u(x,y) = f(x)g(y)$. If $u(x,y)$ is analytic and differentiable continuously with respect to x and y in the domain of interest, then the generalized two-dimensional differential transform of the function $u(x,y)$ is given by

$$U_{\alpha,\beta}(k,h) = \frac{[(D_{x_0}^\alpha)^k(D_{y_0}^\beta)^h u(x,y)]_{(x_0,y_0)}}{\Gamma(\alpha k + 1)\Gamma(\beta h + 1)} \qquad (3)$$

where $0 < \alpha, \beta \leq 1$.

The generalized differential inverse transform of $U_{\alpha,\beta}(k,h)$ is defined as

$$u(x,y) = \sum_{k=0}^{\infty}\sum_{h=0}^{\infty} U_{\alpha,\beta}(k,h)(x - x_0)^{k\alpha}(y - y_0)^{h\beta} \qquad (4)$$

Let $U_{\alpha,\beta}(k,h), V_{\alpha,\beta}(k,h)$ and $W_{\alpha,\beta}(k,h)$ be the differential transform of $u(x,y), v(x,y)$ and $w(x,y)$ respectively.

Property 2.1. If $w(x,y) = c_1 u(x,y) \pm c_2 v(x,y)$, then $W_{\alpha,\beta}(k,h) = c_1 U_{\alpha,\beta}(k,h) \pm c_2 V_{\alpha,\beta}(k,h)$, where c_1, c_2 are constants.

Property 2.2. If $w(x,y) = (D_{x_0}^\alpha)^m v(x,y)$ where $m - 1 < \alpha \leq m$, $m \,\epsilon\, N$ then $W_{\alpha,\beta}(k,h) = \frac{\Gamma(\alpha(k+m)+1)}{\Gamma(\alpha k+1)} V_{\alpha,\beta}(k+m,h)$

Property 2.3. If $w(x,y) = (D_{y_0}^\beta)^n v(x,y)$ where $n - 1 < \beta \leq n$, $n \,\epsilon\, N$ then $W_{\alpha,\beta}(k,h) = \frac{\Gamma(\beta(h+n)+1)}{\Gamma(\beta h+1)} V_{\alpha,\beta}(k,h+n)$

3   Mathematical Modeling

In order to investigate the temperature distribution pattern in a three fluid cross flow heat exchanger, the governing equations describing the energy transfer among the fluid streams are constructed with the following assumptions.

- All the fluids are assumed to be at steady state.
- Exchanger is perfectly insulated from the environment. Therefore, the heat transfer occurs between hot and cold fluids only.
- All physical parameters including the mass flow rates of incoming fluids are constant.
- Axial conduction in the fluid is neglected.
- No heat source and sink is considered inside the fluid steams.

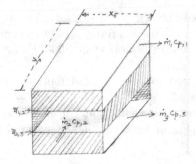

Fig. 1. Schematic representation of a single pass cross flow three fluid plate fin type of heat exchanger

Under these assumptions (Noel et al. 1968), the steady state energy equations describing the two dimensional temperature distribution of a three fluid cross flow heat exchanger (Fig. 1) with Caputo type space fractional derivative are obtained as a system of fractional partial differential equations as follows.

$$\frac{\partial^\alpha \theta_1}{\partial x^\alpha} = \frac{u_{1,2}Y_0}{\dot{m}_1 C_{p,1}}(\theta_2 - \theta_1) \tag{5}$$

$$\frac{\partial^\beta \theta_2}{\partial y^\beta} = \frac{u_{1,2}X_0}{\dot{m}_2 C_{p,2}}(\theta_1 - \theta_2) + \frac{u_{2,3}X_0}{\dot{m}_2 C_{p,2}}(\theta_3 - \theta_2) \tag{6}$$

$$\frac{\partial^\alpha \theta_3}{\partial x^\alpha} = \frac{u_{2,3}Y_0}{\dot{m}_3 C_{p,3}}(\theta_2 - \theta_3) \tag{7}$$

where, $0 < \alpha, \beta \leq 1$. The inlet boundary conditions are $\theta_1(0, y) = \theta_{1i}$, $\theta_2(x, 0) = \theta_{2i}$ and $\theta_3(0, y) = \theta_{3i}$.

Assume $P = \frac{u_{1,2}Y_0}{\dot{m}_1 C_{p,1}}$, $Q = \frac{u_{1,2}X_0}{\dot{m}_2 C_{p,2}}$, $R = \frac{u_{2,3}X_0}{\dot{m}_2 C_{p,2}}$, $S = \frac{u_{2,3}Y_0}{\dot{m}_3 C_{p,3}}$ as constants.

4   Fractional Differential Transform Method

Using fractional differential transform method the Eqs. (5, 6, 7) and the boundary conditions are reduced to the following recurrence relations.

$$\frac{\Gamma(\alpha(k+1)+1)}{\Gamma(\alpha k + 1)}U_{\alpha,\beta}(k+1, h) = P(V_{\alpha,\beta}(k, h) - U_{\alpha,\beta}(k, h)) \tag{8}$$

$$\frac{\Gamma(\beta(h+1)+1)}{\Gamma(\beta h + 1)}V_{\alpha,\beta}(k, h+1) = Q(U_{\alpha,\beta}(k, h) - V_{\alpha,\beta}(k, h)) \tag{9}$$

$$+R(W_{\alpha,\beta}(k, h) - V_{\alpha,\beta}(k, h))$$

$$\frac{\Gamma(\alpha(k+1)+1)}{\Gamma(\alpha k + 1)}W_{\alpha,\beta}(k+1, h) = S(V_{\alpha,\beta}(k, h) - W_{\alpha,\beta}(k, h)) \tag{10}$$

and

$$U(0,h) = \theta_{1i}\delta(h) = \begin{cases} \theta_{1i}, & h = 0 \\ 0, & \text{otherwise.} \end{cases} \quad (11)$$

$$V(k,0) = \theta_{2i}\delta(k) = \begin{cases} \theta_{2i}, & k = 0 \\ 0, & \text{otherwise.} \end{cases} \quad (12)$$

$$W(0,h) = \theta_{3i}\delta(h) = \begin{cases} \theta_{3i}, & h = 0 \\ 0, & \text{otherwise.} \end{cases} \quad (13)$$

where, $U_{\alpha,\beta}(k,h)$, $V_{\alpha,\beta}(k,h)$ and $W_{\alpha,\beta}(k,h)$ are the differential transform of $\theta_1(x,y)$, $\theta_2(x,y)$ and $\theta_3(x,y)$ respectively.

Solving the recurrence relations and applying the inverse fractional differential transformation (4) the solution of the given problem is obtained as a power series as shown below.

$$\theta_1(x,y) = \sum_{k=0}^{m}\sum_{h=0}^{n} U_{\alpha,\beta}(k,h)(x-x_0)^{k\alpha}(y-y_0)^{h\beta} \quad (14)$$

and

$$\theta_2(x,y) = \sum_{k=0}^{m}\sum_{h=0}^{n} V_{\alpha,\beta}(k,h)(x-x_0)^{k\alpha}(y-y_0)^{h\beta} \quad (15)$$

$$\theta_3(x,y) = \sum_{k=0}^{m}\sum_{h=0}^{n} W_{\alpha,\beta}(k,h)(x-x_0)^{k\alpha}(y-y_0)^{h\beta} \quad (16)$$

To validate our result, we consider the heat exchanger (Fig. 1) with parameters $P = 0.25$, $Q = 0.50$, $R = 0.0625$, $S = 0.125$, $X = 1$, $Y = 5$ units, $\theta_1(0,y) = \theta_{1i} = 300$, $\theta_2(x,0) = \theta_{2i} = 100$ and $\theta_3(0,y) = \theta_{3i} = 500$ as in (Noel et al. 1968).

In order to verify the convergence of the series solutions obtained in the Eqs. (14–16), the sequence of errors are collected by increasing the number of terms in the series and is proved as a Cauchy sequence (Bagyalakshmi et al. 2016). Table 1 reveals the convergence of the series solution.

The first few terms $U_{\alpha,\beta}(k,h)$, $V_{\alpha,\beta}(k,h)$ and $W_{\alpha,\beta}(k,h)$ of the series solutions (14–16) are listed below.

$$U_{\alpha,\beta}(0,0) = \theta_{1i}, U_{\alpha,\beta}(0,h) = 0 \text{ for } h \neq 0, \quad (17)$$

$$U_{\alpha,\beta}(1,0) = \frac{P}{\Gamma(\alpha+1)}[\theta_{2i} - \theta_{1i}], \quad (18)$$

$$U_{\alpha,\beta}(1,1) = \frac{P}{\Gamma(\alpha+1)}\left[\frac{1}{\Gamma(\beta+1)}[Q(\theta_{1i} - \theta_{2i}) + R(\theta_{3i} - \theta_{2i})]\right], \quad (19)$$

$$V_{\alpha,\beta}(0,0) = \theta_{2i}, V_{\alpha,\beta}(k,0) = 0 \text{ for } k \neq 0, \tag{20}$$

$$V_{\alpha,\beta}(0,1) = \frac{1}{\Gamma(\beta+1)} \left[Q\left(\theta_{1i} - \theta_{2i}\right) + R\left(\theta_{3i} - \theta_{2i}\right) \right], \tag{21}$$

$$V_{\alpha,\beta}(1,1) = \frac{1}{\Gamma(\beta+1)} \left[Q\left[\frac{P}{\Gamma(\alpha+1)}\left(\theta_{2i} - \theta_{1i}\right) \right] \right]$$
$$+ \frac{1}{\Gamma(\beta+1)} \left[R\left[\frac{S}{\Gamma(\alpha+1)}\left(\theta_{2i} - \theta_{3i}\right) \right] \right], \tag{22}$$

$$W_{\alpha,\beta}(0,0) = \theta_{3i}, W_{\alpha,\beta}(0,h) = 0 \text{ for } h \neq 0, \tag{23}$$

$$W_{\alpha,\beta}(1,0) = \frac{S}{\Gamma(\alpha+1)}\left[\theta_{2i} - \theta_{3i}\right], \tag{24}$$

$$W_{\alpha,\beta}(1,1) = \frac{S}{\Gamma(\alpha+1)} \left[\frac{1}{\Gamma(\beta+1)} \left[Q\left(\theta_{1i} - \theta_{2i}\right) + R\left(\theta_{3i} - \theta_{2i}\right) \right] \right]. \tag{25}$$

Table 1. Error estimate for the hot and cold fluid streams: $\alpha = \beta = 1$

(x_i, y_j)	$Ehot1_{4,4}(x_i, y_j)$	$Ehot1_{7,7}(x_i, y_j)$	$Ehot1_{10,10}(x_i, y_j)$
(0.1356, 4.576)	0.2233	$-7.6934e{-}004$	6.7452e$-$007
(0.322, 3.729)	0.1955	$-4.6361e{-}004$	2.1511e$-$007
(0.5085, 2.966)	0.1112	$-1.6169e{-}004$	3.6119e$-$008
(0.7119, 2.034)	0.0568	2.4123e$-$005	1.4489e$-$009
(0.8644, 0.8475)	0.0316	1.1293e$-$004	$-1.4142e{-}008$
(0.9492, 0.1695)	-0.0093	5.5764e$-$005	$-1.1544e{-}008$
(x_i, y_j)	$Ecold_{4,4}(x_i, y_j)$	$Ecold_{7,7}(x_i, y_j)$	$Ecold_{10,10}(x_i, y_j)$
(0.1356, 4.576)	5.9152	-0.0233	2.0133e$-$005
(0.322, 3.729)	1.7908	-0.0062	2.5545e$-$006
(0.5085, 2.966)	0.4447	-0.0014	1.9724e$-$007
(0.7119, 2.034)	0.0187	$-1.1325e{-}004$	$-3.1200e{-}009$
(0.8644, 0.8475)	-0.0438	3.1833e$-$005	$-3.8470e{-}009$
(0.9492, 0.1695)	-0.0170	9.9409e$-$006	$-1.5311e{-}009$
(x_i, y_j)	$Ehot2_{4,4}(x_i, y_j)$	$Ehot2_{7,7}(x_i, y_j)$	$Ehot2_{10,10}(x_i, y_j)$
(0.1356, 4.576)	0.4388	-0.0015	1.3266e$-$006
(0.322, 3.729)	0.3720	$-8.9285e{-}004$	4.1354e$-$007
(0.5085, 2.966)	0.1850	$-3.0789e{-}004$	6.7757e$-$008
(0.7119, 2.034)	-0.0185	5.6037e$-$005	2.8105e$-$009
(0.8644, 0.8475)	-0.3491	3.4276e$-$004	$-3.6615e{-}008$
(0.9492, 0.1695)	-0.7683	4.9072e$-$004	$-7.6099e{-}008$

5   Results and Discussion

Using inverse FDTM, the values of $U_{\alpha,\beta}(k,h)$, $V_{\alpha,\beta}(k,h)$ and $W_{\alpha,\beta}(k,h)$ are obtained for different values of fractional parameter and Table 2 is presented for $\alpha = \beta = 1$. Using the analytical results the temperature distribution of three fluid streams are predicted for various values of fractional parameter and illustrated graphically from Figs. 2 and 3. The contour plots obtained for various values of fractional order shown in Fig. 4, clearly illustrates the anomalous behaviour in the temperature distribution. From Fig. 4, it is observed that the average temperature distribution obtained in the heat transfer process for various fractional values differs significantly. To validate this result, the statistical analysis (Bagyalakshmi et al. 2017) is carried out among the four group means of temperature distribution using ANOVA at 0.05 level of significance and the results are provided in Table 3. The Fig. 5 clearly depicts the significant difference between the mean temperature distribution for each fluid streams at various fractional values. Hence, the alternate hypothesis stating the four group means of temperature distribution is significantly different is accepted with 0.95 confidence level.

Table 2. Differential transform of $U_{\alpha,\beta}(x,y)$, $V_{\alpha,\beta}(x,y)$, $W_{\alpha,\beta}(x,y)$ for $\alpha = \beta = 1$

$U_{\alpha,\beta}(k,h)$	0	1	2	3	4	5
0	300.0000	0	0	0	0	0
1	−50.0000	15.6250	−1.4648	0.0916	−0.0043	0
2	6.2500	−1.9531	0.5737	−0.0359	0.0017	0
3	−0.5208	0.1628	−0.0478	0.0124	−0.0006	0
4	0.0326	−0.0102	0.0030	−0.0008	0.0002	0
5	−0.0016	0.0005	−0.0001	0.0000	−0.0000	0
$V_{\alpha,\beta}(k,h)$	0	1	2	3	4	5
0	100.0000	62.5000	−5.8594	0.3662	−0.0172	0.0006
1	0	−28.1250	3.1250	−0.1953	0.0092	−0.0003
2	0	6.6406	−1.6113	0.1127	−0.0053	0.0002
3	0	−1.0742	0.2604	−0.0390	0.0020	−0.0001
4	0	0.1322	−0.0320	0.0048	−0.0005	0.0000
5	0	0	0	0	0	0
$W_{\alpha,\beta}(k,h)$	0	1	2	3	4	5
0	500.0000	0	0	0	0	0
1	−200.0000	31.2500	−2.9297	0.1831	−0.0086	0
2	50.0000	−14.8438	1.5137	−0.0946	0.0044	0
3	−8.3333	2.4740	−0.5208	0.0345	−0.0016	0
4	1.0417	−0.3092	0.0651	−0.0092	0.0005	0
5	−0.1042	0.0309	−0.0065	0.0009	−0.0001	0

By reducing the fractional model (5, 6, 7) to the integer order $\alpha = \beta = 1$, the temperature distributions are computed at various locations inside the heat transfer domain using both the fractional differential transform method and the finite difference method and is provided in Table 4. The comparison between the solutions obtained by both the approaches FDTM and FDM shows good agreement as shown in Fig. 6. The exit temperatures of the three fluids in a heat exchanger are estimated using both the fractional differential transform method and the finite difference method and are presented in Table 5.

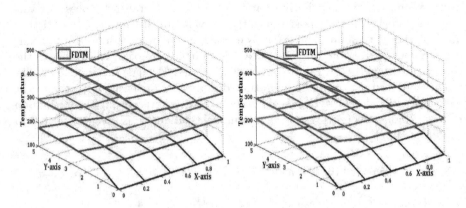

Fig. 2. Temperature distribution of hot and cold fluids for (i) $\alpha = \beta = 0.25, (ii)$ $\alpha = \beta = 0.50$

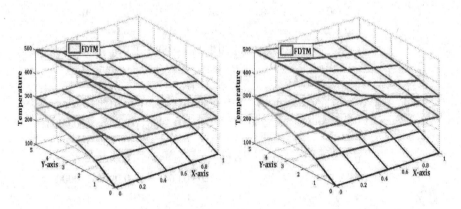

Fig. 3. Temperature distribution of hot and cold fluids for (i) $\alpha = \beta = 0.75, (ii)$ $\alpha = \beta = 1$

Table 3. Results of pairwise mean difference of temperature distribution using t-test

(Hot fluid 1) Fractional parameter	Mean difference	0.95 confidence interval
$\alpha = \beta = 1$ and 0.75	5.0201*	[1.7634, 8.2768]
$\alpha = \beta = 1$ and 0.50	9.9469*	[6.6902, 13.2037]
$\alpha = \beta = 1$ and 0.25	15.7556*	[12.4988, 19.0123]
$\alpha = \beta = 0.75$ and 0.50	4.9268*	[1.6701, 8.1836]
$\alpha = \beta = 0.75$ and 0.25	10.7355*	[7.4787, 13.9922]
$\alpha = \beta = 0.50$ and 0.25	5.8086*	[2.5519, 9.0654]
(Cold fluid) Fractional parameter	Mean difference	0.95 confidence interval
$\alpha = \beta = 1$ and 0.75	17.8601 *	[14.3100, 21.4101]
$\alpha = \beta = 1$ and 0.50	33.7600 *	[30.2100, 37.3101]
$\alpha = \beta = 1$ and 0.25	45.6922*	[42.1421, 49.2422]
$\alpha = \beta = 0.75$ and 0.50	15.8999*	[12.3499, 19.4500]
$\alpha = \beta = 0.75$ and 0.25	27.8321 *	[24.2820, 31.3821]
$\alpha = \beta = 0.50$ and 0.25	11.9321*	[8.3821, 15.4822]
(Hot fluid 2) Fractional parameter	Mean difference	0.95 confidence interval
$\alpha = \beta = 1$ and 0.75	15.2776 *	[9.5766, 20.9785]
$\alpha = \beta = 1$ and 0.50	30.8447*	[25.1437, 36.5457]
$\alpha = \beta = 1$ and 0.25	43.6029 *	[37.9019, 49.3039]
$\alpha = \beta = 0.75$ and 0.50	15.5671*	[9.8662, 21.2681]
$\alpha = \beta = 0.75$ and 0.25	28.3253*	[22.6244, 34.0263]
$\alpha = \beta = 0.50$ and 0.25	12.7582*	[7.0572, 18.4592]

(*Mean difference is statistically significant at $p < 0.05$ level of significance)

Table 4. Comparison of the FDTM with FDM at different space points

Mesh length (60 × 60)	FDTM			FDM			Error
(x, y)	Hot1 air	Hot2 air	Cold air	Hot1 air	Hot2 air	Cold air	In %
(0.1356, 4.576)	486	299.8	281.4	486.1	299.5	279.7	< 3
(0.322, 3.729)	463.6	298	245.8	464.1	296.6	245.8	
(0.5085, 2.966)	438.3	293.8	213.4	439.6	291.5	217.4	
(0.7119, 2.034)	406.4	285.3	176.5	407.7	282.1	180.4	
(0.8644, 0.8475)	373.2	270.8	133.2	374.4	269.1	134	
(0.9492, 0.1695)	351.9	260	106.8	353.4	260	106.8	

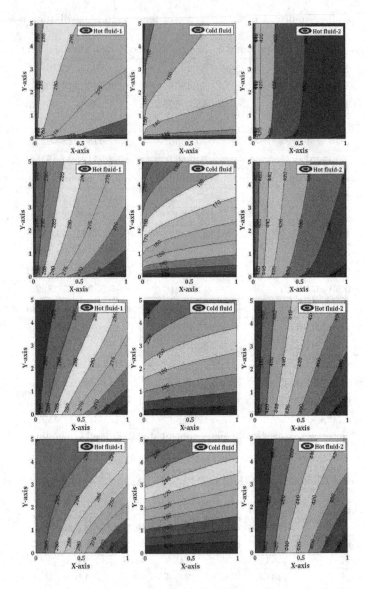

Fig. 4. Contour plots of the temperature profile of the cold and the hot fluid streams for various fractional parameter $\alpha = \beta = 0.25,\ 0.50,\ 0.75,\ 1$

Table 5. Observed exit temperatures in FDTM and FDM

	FDTM	FDM
Hot fluid ($\theta_{1,out}$)	255.8 °C	256.3 °C
Cold fluid ($\theta_{2,out}$)	302.8 °C	301.7 °C
Hot fluid ($\theta_{3,out}$)	342.6 °C	344.1 °C

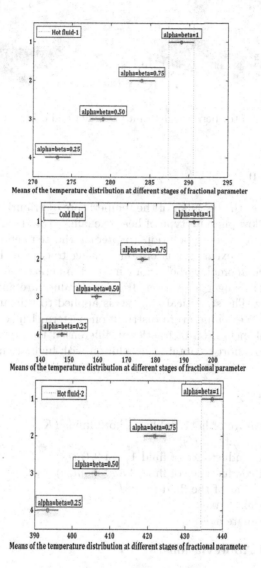

Fig. 5. Multi comparison among the four group means of temperature distribution using t test: (a) Hot fluid stream 1, (b) Cold fluid stream, (c) hot fluid stream 2

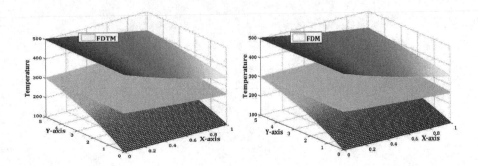

Fig. 6. Comparison of temperature distribution of hot and cold fluids between FDTM and FDM

6   Conclusion

This paper analyses the behavior of the temperature distribution in a three fluid single pass cross flow plate fin type of heat exchanger through fractional derivatives. The new power series solution to predict the temperature distribution in a three fluid heat exchanger is obtained using fractional differential transform method. The anomalous behaviour in the temperature distributions for a three fluid heat exchanger for various fractional values are analyzed and illustrated graphically. The statistical analysis is applied to validate the significant difference among the temperature distribution pattern. Unlike finite difference method, the implementation of fractional differential transform method needs less computational effort to obtain the solution with high accuracy.

Nomenculture

$\theta_1, \theta_2, \theta_3$ Temperature of hot1, cold and hot2 fluids (K)
$\dot{m}$ Mass flow rate (kg/s)
$u_{1,2}$ Overall conductance of fluid 1 and 2 (m/s)
$u_{2,3}$ Overall conductance of fluid 2 and 3 (m/s)
$c_{p,i}$ Specific heat of the fluid (J/kgK)
X X-coordinate axis
Y Y-coordinate axis

Subscripts and Abbreviations:

FDTM Fractional Differential Transform Method
FDM Finite Difference Method

References

Bagyalakshmi, M., Madhu, G., SaiSundarakrishnan, G.: Explicit solution to predict the temperature distribution and exit temperatures in a heat exchanger using Differential Transform Method. Arab. J. Sci. Eng. **41**, 1825–1834 (2016)

Bagyalakshmi, M., Saisundarakrishnan, G., Madhu, G.: On chaotic behavior of temperature distribution in a heat exchanger. Int. J. Bifurcat. Chaos **27**, 19 pages (2017)

Bildik, N., Konuralp, A., Bek, F., Kucukarslan, S.: Solution of different type of the partial differential equation by differential transform method and Adomians decomposition method. Appl. Math. Comput. **172**, 551–567 (2006)

Chen, H., Shujuan, L., Chen, W.: Finite difference, spectral approximations for the distributed order time fractional reaction diffusion equation on an unbounded domain. J. Comput. Phys. **315**, 84–97 (2016)

Erturk, V., Momani, S.: Solving systems of fractional differential equations using differential transform method. J. Comput. Appl. Math. **215**, 142–151 (2008)

Feng, L.B., Zhuang, P., Liu, F., Turner, I., Gu, Y.T.: Finite element method for space-time fractional diffusion equation. Numer. Algorithms **72**, 749–767 (2016)

Goyal, M., Chakravarty, A., Atrey, M.D.: Two dimensional model for multistream plate fin heat exchangers. Cryogenics **61**, 70–78 (2014)

Kukla, S., Siedlecka, U.: An analytical solution to the problem of time-fractional heat conduction in a composite sphere. Bull. Pol. Acad. Sci. Tech. Sci. **65**, 179–186 (2017)

Liu, Y., Dua, Y., Li, H., Li, J., He, S.: A two-grid mixed finite element method for a nonlinear fourth-order reaction diffusion problem with time-fractional derivative. Comput. Math. Appl. **70**, 2474–2492 (2015)

Mishra, M., Sahoo, P.K.: The effect of temperature nonuniformities on transient behaviour of three-fluid crossflow heat exchanger. Eng. Lett. **18**, 297–302 (2010)

Willis Jr., N.C.: Analysis of three fluid crossflow heat exchangers. J. Heat Transf. **90**, 333–338 (1968). National Aeronautics and Space Administration

Podlubny, I.: Fractional Differential Equations. Academic Press, New York (1999)

Saraireh, M.A.: Simulation of steady-state and dynamic behaviour of a plate heat exchanger. J. Energy Power Eng. **10**, 555–560 (2016)

Shrivastava, D., Ameel, T.A.: Three-fluid heat exchangers with three thermal communications. Part A: general mathematical model. Int. J. Heat Mass Transf. **47**, 3855–3865 (2004)

Zhou, J.K.: Differential Transformation and Its Applications for Electrical Circuits. Huazhong University Press, Wuhan (1986)

Zhua, W., Xie, X., Yang, H., Li, L., Gong, L.: Analytical study on multi-stream heat exchanger include longitudinal heat conduction and parasitic heat loads. Phys. Procedia **67**, 667–674 (2015)

Intensive Analysis of Sub Synchronous Resonance in a DFIG Based Wind Energy Conversion System (WECS) Connected with Smart Grid

S. Arockiaraj[✉], B. V. Manikandan, and B. Sakthisudharsun

Mepco Schlenk Engineering College, Sivakasi, Tamil Nadu, India
arockiarocks@gmail.com

Abstract. The advancement in Doubly-Fed induction generator (DFIG) deployed in fields can produce enormous power irrespective of wind speed. Since the cost of DFIG is very high, protection of machines components against super and sub synchronous oscillations must be assured in any operating condition. In this paper the causes and effect for the resonant conditions are well addressed. The DFIG model is developed using MATLAB software and mathematical equations are derived for the considered IEEE first benchmark system. The Eigen value analysis is performed and it's validated with help of PSCAD/EMTDC simulation for various levels of series compensation and wind power penetration. The impact of sub synchronous resonance (SSR) is clearly identified and the corresponding series compensation and power level are measured. The induction generator effect and torsional interaction are the major issues for the SSR in DFIG. These two effects are well analyzed and stability limit in terms of series capacitor and power penetration is obtained.

Keywords: Doubly-fed induction generator · Induction generator effect
SSR · Torsional interaction

1 Introduction

Sub synchronous resonance is an electrical power system condition where the electric network exchanges energy with a turbine generator at one or more of the natural frequencies of the combined system below the synchronous frequency of the system. The analysis of SSR is well explained with the impact of series compensated system [1]. The classification of SSR are induction generator effect (IGE), torsional interaction (TI) and torsional amplification [2]. IGE and TI are dealing with steady state phenomenon and torsional amplification related with the transient state of the system. Induction generator effect is the interaction between the generator and electrical network, torsional interaction is the impact of mechanical dynamics introduced by masses of turbines. Wind Energy Conversion System (WECS) using induction generators and the effect of SSR was analyzed in the paper [3]. It was proved that the unstable condition occurred under sub synchronous frequency due to negative resistance at the

© Springer Nature Singapore Pte Ltd. 2018
G. Ganapathi et al. (Eds.): ICC3 2017, CCIS 844, pp. 242–253, 2018.
https://doi.org/10.1007/978-981-13-0716-4_20

terminal of the machine [3]. A paper [4] depicts SSR in a series compensated constant speed wind power production with DFIG.

The time domain simulation was performed to analyze induction generator effect and torsional interaction using PSCAD. However [4] does not interpret the result for small signal analysis. The impact of wind speed variation, series compensation level and parameters of the current controller under SSR is explicated with modeling of DFIG based WECS [5]. This paper does not give the relation between turbine parameters and torsional oscillation modes [5]. The analysis of SSR on DFIG under different operation condition is worthy of saving the masses of the turbine in future with high wind power penetration.

The remaining parts of the paper are summarized as follows. Section 2 elucidates the mathematical modeling of DFIG based wind farm connected to series compensated network. Sections 3 and 4 include the results of small signal analysis and time domain simulation of induction generator effect and torsional oscillations. Section 5 concludes the paper.

2 System Model

The modified IEEE first bench mark (FBM) system is considered as study system. It consists of an aggregated model of doubly fed induction generator based WECS connected to a series compensated transmission line. The study system considered is shown in Fig. 1. In [4], the power rating of the WECS is 746 MW and the transmission voltage level is 500 kV. In [10], the power rating of the WECS is scaled down to 100 MW and the voltage level of the transmission network is reduced to 132 kV. In this paper, an aggregated (2 MW) DFIG model is taken and the voltage level of the transmission network is 500 kV. The machine and the network parameters are listed in Appendix.

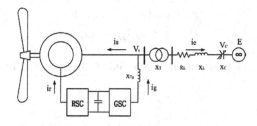

Fig. 1. Study system

2.1 Induction Generator Model

The dynamic model of sixth order [8] is referred for the DFIG with rotor side converter. The model can be represented as

$$\frac{d}{dt}(X_G) = AX_G + BU \tag{1}$$

where

$$X_G = \begin{bmatrix} i_{qs} & i_{ds} & i_{qr} & i_{dr} \end{bmatrix}^T$$
$$U = \begin{bmatrix} v_{qs} & v_{ds} & v_{qr} & v_{dr} \end{bmatrix}^T$$

2.2   Series Compensated Network Model

Modeling of induction machines are based on synchronous reference frame [8]. The same concept is taken for modeling of the network consisting of direct and quadrature axis voltages and current across capacitor. The direct and quadrature axis voltages of bus terminal and infinite bus are derived. The network state variables are derived and described as

$$X_{nw} = \begin{bmatrix} v_{cq} & v_{cd} & i_q & i_d \end{bmatrix}^T$$

2.3   DC Link Model

First order model is referred to formulate the dynamics of capacitor in the dc link between stator side and rotor side converter. It is derived as

$$V_{dc}C\left(\frac{d}{dt}(V_{dc})\right) = P_r - P_g \tag{2}$$

where

$$P_r = 0.5\left(v_{qr}i_{qr} + v_{dr}i_{dr}\right)$$
$$P_g = 0.5\left(v_{qg}i_{qg} + v_{dg}i_{dg}\right)$$

P_r, P_g are the active power at the Rotor Side Converter and Grid Side Converter.

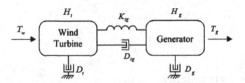

Fig. 2. Two-mass drive train system

2.4   Torsional Dynamics Model

The two-mass drive train system is widely used for studying the power system stability including the wind turbine. A typical two-mass drive train system is shown in Fig. 2.

The following first order differential equations (3)–(6) are describing the behaviour of two mass drive train system. The load angle, speed variation for generator and turbine are derived as follows

$$\frac{d}{dt}(\delta_t) = \omega_t - \omega_0 \tag{3}$$

$$\frac{d}{dt}(\delta_g) = \omega_g - \omega_0 \tag{4}$$

$$2H_t\frac{d}{dt}(\omega_t) = T_t - K_{tg}(\delta_t - \delta_g) - D_{tg}(\omega_t - \omega_g) - D_t\omega_t \tag{5}$$

$$2H_g\frac{d}{dt}(\omega_g) = K_{tg}(\delta_t - \delta_g) + D_{tg}(\omega_t - \omega_g) - D_g\omega_g - T_g \tag{6}$$

Where,

δ_{tg} Torsional angle between wind turbine and generator (rad)
ω_t Angular speed of wind turbine (rad/s)
ω_g Angular speed of generator (rad/s)
H_t Inertia constant of wind turbine (s)
H_g Inertia constant of generator (s)
D_t Damping coefficient of wind turbine (pu)
D_g Damping coefficient of generator (pu)
D_{tg} Damping coefficient between wind turbine and generator (pu)
K_{tg} Shaft stiffness between wind turbine and generator (pu)
T_w Mechanical torque input to wind turbine (pu)
T_g Electromagnetic torque output of generator (pu)

2.5   DFIG Converter Controls

The combined control loops for the rotor side converter and grid side converter have been modeled in this paper. The control strategies for the loops are referred as in [9]. The reference torque is derived from the lookup table. This look up can be obtained from Table 1 given below. When wind speed is more than the rated speed, it is a

Table 1. Rotor shaft speed and mechanical power loop

v_{wind} (m/s)	7	8	9	10	11	12
ω_m		0.75	0.85	0.95	1.05	1.15	1.25
P_m		0.32	0.49	0.69	0.95	1.25	1.60
T_m		0.43	0.58	0.73	0.9	1.09	1.28

constant value. When wind speed is less than the rated speed, the reference torque is the optimal torque corresponding to the measured rotating speed. Using this lookup table, maximum wind power can be obtained from wind turbine by tuning gain values of controller.

3 Results of Eigen Value Analysis for Various Operating Levels

Eigen value analysis is used to find stability of the DFIG for small disturbance. Eigen values are calculated for the IEEE first benchmark system (Fig. 1) with different operating conditions. The various operating conditions considered are:

1. Compensation level is fixed and wind power penetration is varied from 100 MW to 500 MW.
2. The wind penetration level is fixed and series compensation level is varied between 70% and 90%.

For these operating conditions, Eigen value analysis is carried out and the results are presented in Tables 2 and 3.

Table 2. System eigen values for different size of wind WECS with series compensation level (K).

Modes	Wind power penetration		
	100 MW	200 MW	300 MW
K = 70%			
Network Mode-1	−6.523 ± 2768.8i	−7.079 ± 2304.9i	−7.492 ± 2221.7i
Network Mode-2	−7.998 ± 2014.9i	−7.994 ± 1551i	−8.766 ± 1368.9i
Sup. Sync. Mode	−6.169 ± 546.63i	−6.142 ± 587.73i	−6.067 ± 640.03i
Electrical Mode	−2.009 ± 206.04i	−1.424 ± 164.24i	−1.353 ± 141.45i
Elect.mech.Mode	−4.788 ± 41.938i	−4.356 ± 40.321i	−3.996 ± 38.883i
Torsional Mode	−0.347 ± 3.6326i	−0.350 ± 3.6222i	−.3540 ± 3.6122i
System. Mode	−9.18	−8.175	−7.33
K = 75%			
NM-1	−6.234 ± 2532.4i	−7.0473 ± 2231.4i	−7.284 ± 1957.4i
NM-2	−7.6896 ± 1777.4i	−7.5896 ± 1477.4i	−8.61 ± 1203.6i
SuperSync.Mode	−6.8356 ± 549.34i	−6.8356 ± 588.34i	−6.957 ± 634.27i
Electrical Mode	−1.3286 ± 211.73i	−1.3286 ± 161.73i	−1.1892 ± 126.64i
Elect.mech.Mode	−5.529 ± 46.122i	−5.539 ± 36.122i	−3.996 ± 39.372i
Torsional Mode	−0.3543 ± 3.5856i	−0.5873 ± 3.585i	−0.335 ± 3.593i
Stability Mode	−9.7762	−8.7735	−7.876

(continued)

Table 2. (*continued*)

Modes	Wind power penetration		
	100 MW	200 MW	300 MW
K = 80%			
NM-1	−6.1624 ± 2531.4i	−7.0173 ± 2211.4i	−7.165 ± 1955.1i
NM-2	−7.5986 ± 1777.4i	−7.4216 ± 1466.4i	−8.41 ± 1201.4i
SuperSync.Mode	−6.8456 ± 561.34i	−6.8317 ± 589.34i	−6.973 ± 654.16i
Electrical Mode	−1.3456 ± 211.73i	−1.3116 ± 160.3i	−1.168 ± 113.12i
Elect.mech.Mode	−5.589 ± 46.122i	−5.529 ± 36.122i	−3.997 ± 39.375i
Torsional Mode	0.35673 ± 3.583i	−0.5773 ± 3.5856i	−0.325 ± 3.581i
Stability Mode	−9.7754	−9.1635	−8.217
K = 85%			
NM-1	−6.117 ± 2472.5i	−7.0184 ± 2057.4i	−7.180 ± 1927.5i
NM-2	−7.587 ± 1818.5i	−7.345 ± 1203.6i	−8.342 ± 1173.6i
SuperSync.Mode	−6.938 ± 563.69i	−6.967 ± 594.27i	−6.987 ± 663.39i
Electrical Mode	−1.361 ± 187.59i	−1.267 ± 116.64i	**0.249 ± 113.45i**
Elect.mech.Mode	−5.675 ± 53.139i	−3.478 ± 36.372i	−3.998 ± 40.314i
Torsional Mode	−0.356 ± 3.5483i	−0.387 ± 3.593i	−0.323 ± 3.529i
Stability Mode	−9.796	−9.398	−8.655
K = 90%			
NM-1	−6.69 ± 2463.5i	−7.0128 ± 2049i	−7.1672 ± 1916.8i
NM-2	−7.2674 ± 1718.6i	−7.1687 ± 1185.1i	−8.3107 ± 1133i
SuperSync.Mode	−6.9455 ± 567.15i	−6.9514 ± 612.26i	−6.993 ± 678.23i
Electrical Mode	−0.9894 ± 195.43i	0.23908 ± 108.3i	0.95964 ± 131.5i
Elect.mech.Mode	−6.156 ± 53.636i	−5.9443 ± 41.169i	−4.0379 ± 40.79i
Torsional Mode	−0.341 ± 3.5231i	−0.47541 ± 3.675i	−0.32539 ± 3.42i
Stability Mode	−10.978	−10.589	−10.346

Table 3. System eigen values for 400&500 MW size of WECS with series compensation level (K).

Modes	400 MW	500 MW
K = 70%		
NM-1	−7.629 ± 2021.5i	−7.794 ± 1897.4i
NM-2	−9.252 ± 1267.7i	−9.53 ± 1212.6i
SuperSync.Mode	−5.926 ± 624.3i	−5.936 ± 634.33i
Electrical Mode	−1.281 ± 126.84i	−1.382 ± 126.64i
Elect.mech.Mode	−3.691 ± 37.575i	−3.426 ± 36.472i
Torsional Mode	−0.368 ± 3.6024i	−0.375 ± 3.693i
Stability Mode	−6.611	−5.989

(*continued*)

Table 3. (*continued*)

Modes	400 MW	500 MW
K = 75%		
NM-1	−7.494 ± 1967.4i	−7.778 ± 1927.1i
NM-2	−9.13 ± 1213.6i	−9.23 ± 1202.4i
SuperSync.Mode	−5.936 ± 635.27i	−5.977 ± 651.13i
Electrical Mode	−1.176 ± 116.78i	−1.178 ± 117.34i
Elect.mech.Mode	−3.717 ± 38.389i	−3.586 ± 37.282i
Torsional Mode	−0.345 ± 3.673i	−0.365 ± 3.433i
Stability Mode	−6.984	−5.986
K = 80%		
NM-1	−7.794 ± 1957.4i	−7.754 ± 1917.4i
NM-2	−9.52 ± 1213.6i	−9.16 ± 1201.1i
SuperSync.Mode	−5.936 ± 634.27i	−5.986 ± 664.27i
Electrical Mode	**0.0926 ± 115.2i**	**0.132 ± 115.64i**
Elect.mech.Mode	−3.826 ± 39.372i	−3.716 ± 38.372i
Torsional Mode	−0.344 ± 3.596i	−0.360 ± 3.413i
Stability Mode	−7.586	−6.356
K = 85%		
NM-1	−7.618 ± 1928.9i	−7.174 ± 1901i
NM-2	−9.133 ± 1275.1i	−9.087 ± 1200.2i
SuperSync.Mode	−6.049 ± 649.1i	−5.994 ± 669.11i
Electrical Mode	**0.6187 ± 112.63i**	**0.8924 ± 87.56i**
Elect.mech.Mode	−3.996 ± 39.499i	−4.815 ± 41.997i
Torsional Mode	−0.343 ± 3.5243i	−0.342 ± 3.3197i
Stability Mode	−8.1976	−7.743
K = 90%		
NM-1	−7.0119 ± 1863i	−7.0166 ± 1903.7i
NM-2	−9.049 ± 1109.2i	−8.422 ± 1059.9i
SuperSync.Mode	−6.3456 ± 649.45i	−6.2314 ± 695.91i
Electrical Mode	**1.4562 ± 106.51i**	**1.7598 ± 84.297i**
Elect.mech.Mode	−3.9564 ± 40.486i	−5.7522 ± 42.243i
Torsional Mode	−0.3473 ± 3.507i	−0.3374 ± 3.2531i
Stability Mode	−9.787	−9.2934

NM-Network Mode

It is observed from Table 2 that the electrical mode is most sensitive for the variation of series compensation level and the DFIG power penetration. When series compensation level and generator power output increase, the electrical mode generally becomes unstable. The electrical mode becomes unstable for 200 MW power output at 90% series compensation denotes induction generator effect induced in the study system. The electrical mode becomes unstable at 85% compensation with 300 MW WECS output, whereas for 400 and 500 MW, the system becomes unstable at 80% compensation level.

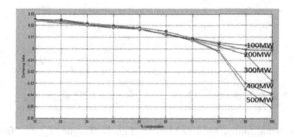

Fig. 3. Simulation result for Eigen value analysis

The damping ratio of various operating conditions are summarized and shown in Fig. 3. The damping ratio of electrical mode is decreasing while increasing the compensation level as well as wind power penetration level. Steady-state SSR of the system for various operating conditions is analyzed and stable/unstable condition of study system is presented in Table 4. The system goes to unstable due to the induction generator effect.

Table 4. Analysis of steady state SSR for various operating levels

K	100 MW	200 MW	300 MW	400 MW	500 MW
70%	S	S	S	S	S
75%	S	S	S	S	S
80%	S	S	US	US	US
85%	S	S	US	US	US
90%	S	US	US	US	US

S-Stable US-Unstable K-percentage of compensation

In the Table 2 the torsional modes are stable for all operating conditions. There is a very less impact on the torsional modes. So, the torsional interaction effect is not occurring in the series compensated WECS. From Fig. 3 it is determined that the stability of electrical mode and the frequency of damping decrease with an increase of the series compensation in the network. The resonant frequency of electrical mode reduces severely with an increase in size of the WECS. It is observed that at particular series compensation level the real part of the electrical mode Eigen values becomes positive which leads to unstable state of the system.

4   Simulation Results

4.1   Steady State Effect of Sub Synchronous Resonance

The induction generator effect and torsional interaction are referred to steady-state sub synchronous resonance. The chance of SSR in steady state is identified from the Eigen value analysis presented in Table 2. The effect of induction generator effect are

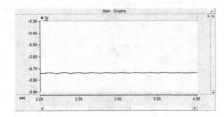

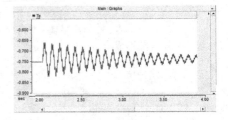

Fig. 4. Electromagnetic torque for 100 MW, series compensation K = 50%

Fig. 5. Torque (T_e) for 200 MW K = 80%

observed to be high at higher level of series compensation. The time domain analysis for different operating conditions are carried out using PSCAD/EMTDC for various wind power penetrations at different series compensation levels (K) and compared with the Eigen value analysis.

To analyze the steady state SSR, WECS with 100 MW penetrations and 50% series compensation is considered. In this case the wind velocity is changed from 5 m/s to 6 m/s at 2 s. It is shown in Fig. 4 that there is no oscillations are observed. So the system is stable.

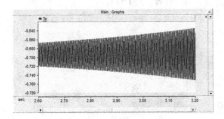

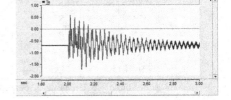

Fig. 6. T_e 400 MW K = 80%

Fig. 7. T_e for 300 MW K = 70%

The response for 200 MW with 80% compensation is shown in Fig. 5. The wind velocity is changed from 5 m/s to 6 m/s at 2 s. Initially the electromagnetic torque of the induction generator is oscillating and it decays after few seconds. In this case the system is marginally stable. In similar manner the analysis for 400 MW wind power penetration with 80% series compensation carried out as follows. Here the steady state disturbance at 2 s was applied by changing wind velocity and the variation of electromagnetic torque for 400 MW system shown in Fig. 6. It emphasis that the critical level of compensation for 400 MW system is 80%. The system is unstable and SSR oscillation grows gradually. This correlation validates the Eigen value analysis.

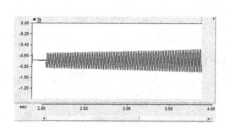

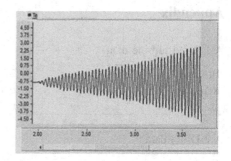

Fig. 8. (T_e) for 300 MW K = 80%   **Fig. 9.** Torque (T_e) for 300 MW K = 85%

4.2 Transient Subsynchronous Resonance

The transient behavior of the study system is analyzed by applying a three phase fault at the receiving end of the transmission line at 2 s and the fault is cleared at 2.1 s.

Case 1: The Wind Power Penetration of 300 MW with 70% Compensation
The response of electromagnetic torque is shown in Fig. 7. It is observed that the oscillations are decreasing and die out after few seconds. So the system is stable in this condition.

Case 2: The Wind Power Penetration of 300 MW with 80% Compensation
Since series compensation increases to 80%, the oscillations of the electromagnetic toque are persisting with almost same amplitude. The oscillations of the electromagnetic torque consist of multiple frequency components. The electromagnetic torque oscillations are shown in Fig. 8. The system is marginally stable in this condition.

Case 3: The Wind Power Penetration of 300 MW with 85% Compensation
In this case the series compensation is increased to 85% and the response is depicted in Fig. 9. The oscillations of electromagnetic torque are growing and the system is unstable in this case. The Eigen value analysis for the assessment of steady-state SSR is carried out for the modified first bench mark model and the results are validated using time-domain simulation in PSCAD/EMTDC software package.

5 Conclusion

In this paper a detailed analysis of SSR in DFIG-based WECS connected to series compensated lines is performed. A comprehensive system is modeled for the study system considered, which is adapted from the modified IEEE First Benchmark System. Eigen value analysis is also performed by using MATLAB to predict the potential of SSR for various levels of WECS and series compensation levels. These results are validated through detailed electromagnetic time domain simulation using PSCAD/EMTDC software. Critical series compensation levels are identified that it could cause steady-state and transient state SSR.

Appendix

A.1 Wind turbine data

Ht	2.5 s
Ksh	0.3 pu/el. rad.
Dsh	0

A.2 DFIG data

Hr	0.5 s
Vs	690 V
Rs	0.00488 pu
Rr	0.00549 pu
Xm	3.95279 pu
Pg	2 MW
ω_s	377 rad./s
Xss	0.09241 pu
Xrs	0.09955 pu
DClink C	9000μF/1200 kV

A.3 AC transmission line data

S_{base}	892.4 MVA
RL	0.02 pu
X2	0.50 pu
Vbase	500 kV
X1	0.14 pu
X3	0.06 pu

References

1. Kundur, P.: Power System Stability and Control. McGraw-Hill, New York (1994)
2. Padiyar, K.P.: Power system dynamics stability and control. IEEE Trans. Power Deliv. **25** (4), 2082 (2010)
3. Limebeer, D., Harley, R.: Sub synchronous resonance of single-cage induction motors. In: IEE Proceedings of Electric Power Applications, vol. 128, no. 1, pt. B, pp. 33–42, January 1981
4. Varma, R.K., Auddy, S., Semsedini, Y.: Mitigation of sub synchronous resonance in a series-compensated wind farm using FACTS controllers. IEEE Trans. Power Del. **23**(3), 1645–1654 (2008)
5. Ostadi, A., Yazdani, A., Varma, R.: Modeling and stability analysis of a DFIG-based wind-power generator interfaced with a series-compensated line. IEEE Trans. Power Del. **24**(3), 1504–1514 (2009)

6. Zhu, C., Fan, L., Hu, M.: Modeling and simulation of a DFIG-based wind turbine for SSR. In: Proceedings of North American Power Symposium (NAPS), October 2009
7. de Jesus, F., Watanabe, E., de Souza, L., Alves, J.: SSR and power oscillation damping
8. Krause, P.: Analysis of Electric Machinery. McGraw-Hill, New York (1986)
9. Mei, F., Pal, B.C.: Modeling of doubly-fed induction generator for power system stability study. In: Proceedings of IEEE Power & Energy General Meeting 2008, July 2008
10. Mohorana, A.K.: Sub synchronous resonance in wind farms, Ph.D. thesis. The University of Western Ontario, Canada (2012)
11. Krause, P.C., Thomas, C.H.: Simulation of symmetrical induction machinery. IEEE Trans. Power Appar. Syst. **84**(11), 1038–1053 (1965)
12. Lubosny, Z.: Wind Turbine Operation in Electric Power System. Springer, Heidelberg (2003)
13. Lara, A., Ekanayake, J., Huges, C.: Wind Energy Generation Modeling and Control. Wiley, Chichester (2009)
14. Fan, L.: Modeling of DFIG based wind farms for SSR analysis (2010)
15. Li, L.: Small signal stability of SCIG based wind farm interconnected system (2010)
16. Louie, K.W.: Aggregating induction motors in a power system based on their standard specifications. In: Proceedings of International Conference on Power System Technology, pp. 1–8, October 2006
17. Louie, K.W.: Aggregating induction motors in a power system based on their standard specifications, October 2006
18. Fan, L., Miao, Z.: SSR Analysis using nyquist stability criterion frequency scan method (Future issue)
19. Limebeer, D.J.N., Harley, R.G.: Sub synchronous resonance of single cage induction motors. IEE Proc. Electr. Power Appl. **128**(1), 33–42 (1981)
20. Ilea, V.: Damping of the poorly damped inter area electro mechanical oscillations by installing FACTS devices (2009)

Author Index

Printed in the United States
By Bookmasters